NOTES ON CRYSTALLINE COHOMOLOGY

by

Pierre Berthelot

Arthur Ogus

Princeton University Press

and

University of Tokyo Press

Princeton, New Jersey

1978

Published in Japan exclusively
by University of Tokyo Press
in other parts of the world by
Princeton University Press

Printed in the United States of America
by Princeton University Press, Princeton, New Jersey

Library of Congress Cataloging in Publication Data will
be found on the last printed page of this book

Contents

Preface

The first seven chapters of these notes reproduce the greatest part of a seminar held by the first author at Princeton University during the spring semester, 1974. The seminar was meant to provide the auditors with the basic tools used in the study of crystalline cohomology of algebraic varieties in positive characteristic, and did not cover all known results on this topic. These notes have the same limited purpose, and should really be considered only as an introduction to the subject.

In Chapter I, we draw a rapid picture of the various cohomology theories for algebraic varieties in characteristic p, and try to explain the specific need for a p-adic cohomology, as well as the motivations of the technical definition of the crystalline site; this chapter is purely introductory, and contains no proof. The second chapter introduces some basic notions of differential calculus, and in particular various presentations of the notions of connection and stratification. Actually, these notions, under this particular form, are relevant to crystalline cohomology only in characteristic zero; but it seemed more convenient to give first the algebraic presentation of ordinary differential calculus, which is more or less familiar to the reader, and then to explain how it has to be modified to yield a good formalism in characteristic p. To do this, we introduce in Chapter III the notion of a divided power ideal. The main result in this chapter is the construction of the divided power envelope of an ideal in an arbitrary commutative ring, which is of constant use in what follows. Chapter IV reviews then the notions of Chapter II with the modifications necessary to work in characteristic p.

With chapter V begins the theory of the crystalline topos. Once we have defined the crystalline site, and described the sheaves on this site, we establish the functoriality of the corresponding topos, and show in particular that if X is an S-scheme, and (I,γ) a divided power ideal in S which extends to X, the crystalline cohomology of X relatively to (S,I,γ) depends only upon the reduction of X modulo I. The chapter ends with a discussion of the relations between the crystalline and the Zariski topoi, which will be used in Chapter VII to relate crystalline and de Rham cohomologies. Chapter VI is devoted to the notion of crystal. First we define crystals, and show how they can be interpreted as modules on a suitable scheme endowed with a quasi-nilpotent integrable connection. We then associate to a complex $K^\bullet$ *of differential operators* of order 1 on a smooth S-scheme Y a complex of crystals, with linear differential, on the crystalline site of any Y-scheme X. In the particular case where $K^\bullet$ is the complex of differential forms on Y relatively to S, we thus obtain a resolution of the structural sheaf of the crystalline topos ("Poincaré lemma").

In Chapter VII, we prove (in a new way) the fundamental property of crystalline cohomology: If X is a closed subscheme of a smooth S-scheme Y (on which p is nilpotent), the crystalline cohomology of X relatively to S is isomorphic to the de Rham cohomology of Y relative to S, with coefficients in the divided power envelope of the ideal of X in Y; more generally, if E is a crystal on X, an analogous result gives the crystalline cohomology of X with coefficients in E. We derive several consequences of this fact, the most important of which is a base changing theorem. As a particular case of this theorem, one gets "universal coefficients" exact sequences, which may be used to relate torsion in crystalline cohomology and "pathologies" in characteristic p, such as varieties having too big de Rham cohomology, non reduced Picard scheme, closed 1-forms not coming from the Albanese variety, etc. Finally, we define crystalline cohomology relative to a p-adically complete noetherian base A, by studying the inverse limit of crystalline cohomologies relatively to the A/p^n.

Chapter VIII, due to the second author, gives an application of these results to Katz's conjecture. Let us briefly recall the conjecture. If X is a proper and smooth variety over a perfect field k, and $n \leq 2\dim(X)$, one associates to X in degree n two convex polygons. The first one, the Hodge polygon, is the convex polygon with sides of slope i and horizontal projection of length $h^i = \dim_k H^{n-i}(X,\Omega^i_{X/k})$. The slopes of the sides of the second one, the Newton polygon, are the slopes of the action of Frobenius on the crystalline cohomology $H^n(X/W(k))$, and the lengths of their horizontal projections are the multiplicities of the corresponding slopes. The conjecture then asserts that the Newton polygon lies above the Hodge polygon, and leads thus to p-adic estimates on the zeroes and poles of the zeta function of X when k is a finite field. It has been proved by Mazur ([6], [7]) when X is projective and has a projective and smooth lifting on W(k), assuming further its Hodge groups have no torsion. Following an idea of Deligne, one first proves a local theorem (which can be regarded as a p-adic version of the Cartier isomorphism), from which one can deduce Katz's conjecture for an arbitrary smooth proper X/k, as well as a stronger conjecture of Mazur [6, p. 663].

Finally, there are two appendices, the first of which gives a rapid sketch of Roby's divided power envelope of a module. The second appendix discusses inverse limits in the context of derived categories, using a method rather different from Houzel's [SGA 5 XV].

Among the topics not discussed in this book, but closely related, we should mention the following:

a) Poincaré duality, the construction of the cohomology class of a cycle, and the Lefschetz fixed point formula, for which the reader may refer to [2];

b) crystalline cohomology of abelian varieties, and its relation to Dieudonné theory, which can be found in Mazur-Messing [8];

c) the slope filtration on crystalline cohomology, and its interpretation, thanks to the De Rham-Witt complex, developed by Bloch and Illusie, following a direction initiated by Mazur (cf. [3], [1], [4], [5]).

Let us finally mention that our knowledge of crystalline cohomology remains far from being as complete as it is for étale cohomology, for example; actually, little progress seems to have been made in most of the questions raised in the introduction of [2].

We hope that our treatment will be comprehensible to anyone with a knowledge of Grothendieck's theory of schemes, i.e. with EGA, as well as the standard facts about algebraic De Rham cohomology. In particular, we have not assumed familiarity with topoi or derived categories, and have tried to provide an informal development (by no means a systematic treatment) of these ideas as we need them.

Apart from some minor details, these notes have been entirely written by Ogus, who tried to recover the rather informal spirit of the seminar; I wish to thank him for all the work he has done. I also thank Princeton University for its hospitality during the spring semester, 1974. Finally, both authors thank Neal Koblitz and Bill Messing for reading the manuscript and making innumerable suggestions, and the typist Ruthie L. Cephas for her patience and precision.

P. Berthelot

References for Preface

[1] M. Artin, B. Mazur, Formal groups arising from algebraic varieties, preprint.

[2] P. Berthelot, Cohomologie cristalline des variétés de caractéristique $p > 0$, Lecture Notes in Math. 407, Springer.

[3] ___________, Slopes of Frobenius on crystalline cohomology, Proc. of Symp. in Pure Math. 29 (1974), p. 315-328.

[4] S. Bloch, Algebraic K-theory and crystalline cohomology, preprint.

[5] L. Illusie, Complexe de de Rham-Witt et cohomologie cristalline, résumé d'un cours à l'Université Paris-Sud, 1976.

[6] B. Mazur, Frobenius and the Hodge filtration I, Bull. A.M.S. 78 n° 5 (1972), p. 653-667.

[7] ________, Frobenius and the Hodge filtration II, Ann. of Math. 98 (1973), p. 58-95.

[8] B. Mazur, W. Messing, Universal extensions and one-dimensional crystalline cohomology, Lecture Notes in Math. 370, Springer.

§1. Introduction. Let k be the field with q elements, X/k a smooth, projective, and geometrically connected scheme. One wants to know how many rational points X has, or more generally, the number c_ν of k_ν-valued points of X where k_ν is the extension of k of degree ν. The values of these numbers are conveniently summarized in the zeta function of X, given by $Z_X(t) = \exp \sum_{\nu=1}^{\infty} \frac{c_\nu}{\nu} t^\nu$. This can also be written as $\prod_x \frac{1}{[1-t^{\deg x}]}$, the product being taken over the closed points x of X, where $\deg x$ means the degree of the residue field $k(x)$ over k. It is clear from the second expression that $Z_X(t)$ is a power series with integral coefficients and constant term 1. The following results represent the culmination of twenty-five years of work by Weil, Dwork, Grothendieck, Deligne, and others.

Theorem I.

1. $Z_X(t)$ is a rational function of t and can be written:

$$Z_X(t) = \frac{P_1(t)\ldots P_{2n-1}(t)}{P_0(t)\ldots P_{2n}(t)}$$

where $n = \dim X$, $P_0(t) = 1-t$, $P_{2n}(t) = 1 - q^n t$, and $P_i(0) = 1$. Moreover $P_i(t)$ is a polynomial of degree β_i, where β_i is the ith Betti-number of a lifting of X to characteristic zero, if one exists. Furthermore, one has a functional equation

$$Z_X\left(\frac{1}{q^n t}\right) = \pm\, q^{n\chi/2}\, t^\chi Z_X(t), \text{ where } \chi = \sum_{i=0}^{2n} (-1)^i \beta_i \ .$$

<u>Theorem II.</u> The polynomials $P_i(t)$ above have integral coefficients. Moreover, if $P_i(t) = \prod_{\alpha=1}^{\beta_i}(1-\omega_\alpha t)$, the complex numbers ω_α have absolute value $q^{i/2}$.

The second theorem was only recently proved by Deligne [2], whereas the first is several years older. It has been well known for some time that Theorem I follows "formally" from the existence of a sufficiently rich cohomological machine ("Weil cohomology"). This is supposed to be (at least) a functor H^* from the category of smooth projective k-schemes X to the category of graded finite dimensional algebras over a field K of characteristic zero, enjoying the following properties:

<u>A</u> i) $H^i(X) = 0$ if $i > 2n$, $i < 0$ $(n = \dim X)$.

ii) There is a canonical map $H^{2n}(X) \longrightarrow K$, ("trace"), an isomorphism if X is geometrically connected.

iii) The multiplication law and trace map induce pairings $H^i(X) \times H^{2n-i}(X) \longrightarrow K$, which are perfect if X is geometrically connected.

<u>B</u> If $\pi_X : X \times Y \to X$ (resp. π_Y) is the canonical map, then π_X^*, π_Y^*, and multiplication induce an isomorphism:

$$H^*(X) \otimes H^*(Y) \to H^*(X \times Y).$$

<u>C</u> Let $C^r(X)$ be the free abelian group generated by the irreducible closed subsets of X of codimension r. There is a natural map ("cycle map")

$$\gamma_X : C^r(X) \to H^{2r}(X)$$

such that:

i) If $f:X \to Y$, and if $f^* : H^*(Y) \to H^*(X)$ is $H^*(f)$ and $f_*:H_*(X) \to H_*(Y)$ is the map induced from $H^*(f)$ by Poincare duality (Aiii), then, wherever f^* and f_* are defined on $C^*(X)$, $f^* \circ \gamma_Y = \gamma_X \circ f^*$ and $\gamma_Y \circ f_* = f_* \circ \gamma_X$.

ii) If $X' \subseteq X$ and $Y' \subseteq Y$ are closed subsets, then $\gamma_X(X') \otimes \gamma_Y(Y') = \gamma_{X \times Y}(X' \times Y')$, where $\otimes$ means via the Kunneth map (B).

iii) If $X = \underline{\mathrm{Spec}}\ k$, we have a commutative diagram:

$$\begin{array}{ccc} C^0(X) & \longrightarrow & H^0(X) \\ \cong \downarrow & & \cong \downarrow \mathrm{tr} \\ \mathbb{Z} & \hookrightarrow & K \end{array} .$$

The above axioms formally imply [8] a Lefschetz fixed point formula: If $f:X \to X$, and if $\Delta \in C^n(X \times X)$ is the diagonal and $\Gamma_f \in C^n(X \times X)$ is the graph of f, then

$$< \Gamma_f, \Delta > = \sum_{i=0}^{2n} (-1)^i \ \mathrm{Tr}\ H^i(f)$$

where the left side means the pairing of (Aiii) applied to $\gamma_{X \times X}(\Gamma_f)$ and $\gamma_{X \times X}(\Delta)$. This formula essentially proves Theorem I, for if we let $F^{\nu}_{X/k}$ be the νth relative Frobenius: $X \to X$, (which is $x \to x^{q^{\nu}}$ in coordinates), $\Delta \cap \Gamma_f$ is easily seen to be a smooth scheme of dimension zero with c_{ν} points. It follows that $c_{\nu} = \Sigma(-1)^i \ \mathrm{Tr}\ H^i(f)$, and elementary manipulations lead quickly to the fundamental formula:

$$Z_X(t) = \prod_{i=0}^{2n} \det[1-t\ H^i(F_{X/k})]^{(-1)^{i+1}} .$$

<u>Note.</u> We can get away with slightly weaker axioms, viz.:

<u>C'</u> i)' The same as above, but γ_X only defined for non-singular cycles.

ii)' Assuming X' and $X'' \subseteq X$ are smooth and intersect transversally, $\gamma_X(X' \cap X'') = \gamma_X(X') \cdot \gamma_X(X'')$.

iii)' If $j: X' \hookrightarrow X$ is a closed immersion and X' is also smooth, $\gamma_X(X') = j_*(1_{X'})$ where $1_{X'} \in H^0(X')$ is the unit element.

iv)' If $x \in X$ is a closed point, $\operatorname{tr} \gamma_X(x) = \deg x$.

These properties A,B,C' (together with a comparison with the cohomology of a lifting, for the statement about the Betti numbers), suffice to prove Theorem I.

In these notes we shall try to explain one such (weak) Weil cohomology - crystalline cohomology. It is denoted H^i_{cris}, and takes its values in the (fraction field of) the ring $W(k)$ of Witt vectors of the ground field k. Let us first briefly review the history of the attempts to build a Weil cohomology, and the special place of the crystalline theory among them.

1) <u>Serre's cohomology of coherent sheaves (FAC).</u>

Of course $H^i(X,F)$ has characteristic p if X does, instead of characteristic zero, but it was hoped that at least one would recover the Betti numbers β_i of a lifting of X from $\beta_i' = \sum_{p+q=i} \dim_k H^q(X,\Omega^p_{X/k})$ Igusa showed that this fails, with an example of a surface X with $\dim H^1(X,\mathcal{O}_X) > \dim \operatorname{Pic}(X)$. This violated the expected $\beta_1 = 2 \dim \operatorname{Pic}(X)$. Later Serre constructed a surface X with $H^1(X,\mathcal{O}_X)$ one dimensional and $H^0(X,\Omega^1_{X/k}) = 0$, so that Alb(X) and therefore also the Picard variety vanish. We now understand that this behavior is due to singularities of the Picard scheme of X. For a further discussion of these and other examples, as well as precise references, we refer the reader Mumford's appendix to Chapter VII of Zariski's book on surfaces [14].

Since we do have the inequalities $\beta'_i \geq \beta_i$, one might try the de Rham hypercohomology $\mathbb{H}^i(X,\Omega^\cdot_{X/k})$. If β''_i is its dimension, we have $\beta'_i \geq \beta''_i$, with inequality caused by the failure of the Hodge $\Rightarrow$ de Rham spectral sequence to degenerate. However for Serre's surface we still have $\beta''_1 = 1 > \beta_1 = 0$ (to see this, convince yourself that Frobenius induces an injection $H^1(X,\mathcal{O}_X) \hookrightarrow \mathbb{H}^1(X,\Omega^\cdot_{X/k})$). If we set $H^i(X) = \bigoplus_{q+p=i} H^q(X,\Omega^p_{X/k})$ or $\mathbb{H}^i(X,\Omega^\cdot_{X/k})$, we do get a (characteristic p) theory satisfying A,B,C', above, and (after some fancy footwork) a congruence formula for Z_X – *but by no means a proof of Theorem I.*

Note. The proof of the congruence formula alluded to above is due to Deligne and should appear some day in SGA. On the other hand, Katz has given a proof in [SGA VII, exp. XII], based on the theory of Dwork.

2) Serre's Witt vector cohomology.

If k is a perfect field of characteristic p, a classical construction yields a canonical lifting of k to a discrete valuation ring $W(k) = \varprojlim W_n(k)$. Serre generalized this to $W(A) = \varprojlim W_n(A)$ for any ring A of characteristic $p > 0$, obtained a sheaf of rings $W_n = W_n(\mathcal{O}_X)$, and then defined $H^i(X,W)$ to be $\varprojlim H^i(X,W_n)$. This is a module over $W(k)$ and hence yields a characteristic zero theory; moreover $H^1(X,W)$ is free and of finite rank. However, since $H^i(X,W) = 0$ if $i > \dim X$, it is clear that this cannot be all of a Weil cohomology. Furthermore, the rank of $H^1(X,W)$ is $\leq 2 \dim \mathrm{Alb} X$, and the inequality is ordinarily strict – in fact if X is a curve, the rank is $2g-\sigma$, where g is the genus and σ is the p-rank of the Jacobian of X [13]. Finally, if $i \geq 2$, Serre showed that even for abelian varieties $H^i(X,W)$ need not be finitely generated (too much torsion) [12].

Nevertheless, Serre [12] was able to "force" a theory for abelian varieties by defining $L(X) = H^1(X,W) \oplus T_p(X^*) \otimes_{\mathbb{Z}_p} W(k)$, where $T_p(X^*)$ is the Tate module of points of order a power of p on the dual. If $\varphi: X \to X$ is a group homomorphism, one gets an endomorphism $L(\varphi)$ of $L(X)$, and Serre proved that the characteristic polynomial of $L(\varphi)$ agreed with the characteristic polynomial in the ℓ-adic Tate module as studied by Weil. Thus L is a good p-adic H^1 for abelian varieties, and should "agree with" crystalline H^1_{cris}. Artin and Mazur (unpublished) have shed some light on the meaning of $H^i(X,W) \otimes_{W(k)} K$, where K is the fraction field of W(k). They show that $H^i(X,W) \otimes K$ is a quotient of $H^i_{\mathrm{cris}}(X) \otimes K$, and under some hypotheses on X, is the part on which Frobenius acts with slopes in $[0,1)$. (The p-adic value of an eigenvalue of F is called the slope of the eigenvalue – it, unlike the eigenvalue itself, is well-defined.) Recent work of Bloch [1] using higher K-theory to generalize the W-construction should shed more light on the <u>rest</u> of crystalline cohomology.* Let us remark that Witt vector cohomology seems intimately related with p-torsion phenomena, the non-smoothness of <u>Pic</u> , etc.

3) <u>Grothendieck's ℓ-adic cohomology.</u>

So far, this is the only theory which is sufficiently rich to prove Theorem II, and it was the first to prove all of Theorem I. This is not the place to even sketch Grothendieck's magnificent idea. Let us only say that for $\ell \neq p$, one has reasonable cohomology groups $H^i(X,\mathbb{Z}/\ell^n\mathbb{Z})$; then to get the Weil cohomology we let $H^i(X,\mathbb{Q}_\ell) = \varprojlim H^i(\bar{X},\mathbb{Z}/\ell^n\mathbb{Z})\otimes\mathbb{Q}$, where $\bar{X}$ is $X \times_k \bar{k}$ and $\bar{k}$ is an algebraic closure of k. We have killed torsion here, since it is not needed for Weil's conjectures, but it does have meaning, because for each n, $H^i(\bar{X},\mathbb{Z}/\ell^n\mathbb{Z}) \cong H^i(X^{an},\mathbb{Z}/\ell^n\mathbb{Z})$, where X^{an} is the analytic space associated with a lifting of X. This is false if $\ell = p$, and so we have no information about the p-torsion in $H^i(X^{an},\mathbb{Z})$.

4) <u>Dwork's p-adic theory.</u>

Strictly speaking, Dwork did not define a Weil cohomology, but worked with a p-adic Banach space, related (not obviously, but by Katz [6]) to de Rham cohomology. He was the first to prove the rationality statement in Theorem I. His ideas have since been expanded by Washnitzer and Monsky, using their "weak completion" $A^\dagger$ of an algebra A of finite type over a discrete valuation ring V of mixed characteristic. If A_0 is smooth (and satisfies an additional technical, but mild, restriction)

*Bloch's work has now been extended by Deligne and Illusie, who have even removed the K-theory.

over the residue field k of V, they find a smooth lifting $A^{\dagger}$ A_0 to V and develop a fixed point formula using $\mathbb{H}^*(\Omega^{\cdot}_{A^{\dagger}/V})\otimes_V K$. Moreover, $H^*(\Omega^{\cdot}_{A^{\dagger}/V})$ depends only on A_0 if $\nu_{\pi}(p) < p-1$, where π is uniformizing parameter of V. The torsion in $\mathbb{H}^*(\Omega^{\cdot}_{A^{\dagger}/V})$ is uncontrollably large, however, and finite dimensionality is still unknown, even after tensoring with K. The liftable and projective case was worked out in detail by Lubkin, who established all of Theorem I using this theory. Monsky has written a nice introduction to this circle of ideas [11].

5) Grothendieck's approach to p-adic cohomology.

Suppose V is a discrete valuation ring with uniformizing parameter π, satisfying $\nu_{\pi}(p) < p$, let $X'/\underline{\text{Spec}}\ V$ be proper and smooth, and let X be the fiber over the residue field k. We ask:

(1) Is $\mathbb{H}^*(X',\Omega^{\cdot}_{X'/V})$ independent (in a "natural way") of the lifting X' of X ?

(2) If so, can we define these groups without referring to the lifting, and get a theory for non-liftable schemes?

If the answers to these questions are yes, we will have a good p-adic theory. We hope, by the way, that the V-module itself is interesting, so that we have information on p-torsion.

In order to understand Goothendieck's attack on these questions, let us explain it in characteristic zero, where we know that $H^*(X^{an},\mathbb{C})$ is a "good" cohomology. The Zariski topology and even the etale topology, are too coarse to give correct answers when calculating the cohomology of $\mathbb{C}$. However, if $X/\mathbb{C}$ is smooth, one can calculate $H^*(X^{an},\mathbb{C})$ by means of its resolution $\Omega^{\cdot}_{X/C}$. Then using GAGA, one gets

$H^*(X^{an},\mathbb{C}) \xrightarrow{\cong} H^*(X^{an},\Omega^{\cdot}_{X/\mathbb{C}}) \xleftarrow{\cong} H^*(X^{zar},\Omega^{\cdot}_{X/\mathbb{C}})$ (if $X/\mathbb{C}$ is proper – but Grothendieck showed that this hypothesis is unnecessary [5]). This suggests that we can recreate $H^*(X^{an},\mathbb{C})$ algebraically by using a "topology" for which "locally" means "infinitesimally", or "differentiably".

To do this Grothendieck introduces the "infinitesimal site" $\mathrm{Inf}(X/\mathbb{C})$, whose objects are nilpotent (infinitesimal) thickenings $U \hookrightarrow T$ of Zariski open subsets U of X, and whose morphisms are commutative diagrams:

$$\begin{array}{ccc} U & \hookrightarrow & T \\ {\scriptstyle incl}\downarrow & & \downarrow{\scriptstyle u} \\ U' & \hookrightarrow & T' \end{array}$$

where the map $u:T \to T'$ is any map making the diagram commute. The "infinitesimal topos" $(X/\mathbb{C})_{inf}$ will be the category of sheaves on $\mathrm{Inf}(X/\mathbb{C})$, and can be described as follows: Its <u>objects</u> are collections $\{F_{(U,T)}$ for each thickening $U \hookrightarrow T\}$, where $F_{(U,T)}$ is a Zariski sheaf on T, together with a collection of maps: $\rho_u : u^*F_{(U',T')} \to F_{(U,T)}$, for each morphism u in $\mathrm{Inf}(X/\mathbb{C})$, satisfying the transitivity condition: If $v:T' \to T''$ is another morphism, the diagram below commutes:

$$\begin{array}{ccc} u^*v^*F_{(U'',T'')} & \xrightarrow{\rho_u^*(\rho_v)} & u^*F_{(U',T')} \\ \cong\downarrow & & \downarrow\rho_u \\ (v\circ u)^*F_{(U'',T'')} & \xrightarrow{\rho_{v\circ u}} & F_{(U,T)} \end{array}$$

A <u>morphism</u> of $(X/\mathbb{C})_{inf}, \varphi : F \to G$, is a collection of maps: $F_{(U,T)} \xrightarrow{\varphi_{(U,T)}} G_{(U,T)}$ making the diagram below commute:

$$\begin{array}{ccc} F_{(U,T)} & \xrightarrow{\varphi_{(U,T)}} & G_{(U,T)} \\ \uparrow & & \uparrow \\ u^*F_{(U',T')} & \xrightarrow{u^*(\varphi_{(U',T')})} & u^*G_{(U',T')} \end{array}$$

for any morphism u in $\mathrm{Inf}(X/\mathbb{C})$. We have already failed to resist an abuse of language we shall hereafter frequently indulge in, namely we shall let "T" stand for the closed immersion $U \longrightarrow T$ and shall write F_T for $F_{(U,T)}$. (This unfortunately obscures the effect of automorphisms of T.)

An important object of $(X/\mathbb{C})_{inf}$ is $\mathcal{O}_{X/\mathbb{C}}$, the "structure sheaf", which to T assigns the sheaf $\mathcal{O}_T$. Less important is the assignment $T \mapsto \mathcal{O}_U$.

If F is an object of $(X/\mathbb{C})_{inf}$, we define $\Gamma(X/\mathbb{C},F)$ to be the set of all assignments $s_T \in \Gamma(T,F_T)$, for each T of $\mathrm{Inf}(X/\mathbb{C})$, such that for any morphism u, $\rho_u : F_{T'} \to u_*F_T$ maps $S_{T'}$ to S_T. Because there are so many objects T and so many maps u, *it is hard to find elements of* $\Gamma(X/\mathbb{C},F)$. (For example, if $X/\mathbb{C}$ is a scheme of finite type, $\Gamma(X,\mathcal{O}_{X/\mathbb{C}}) \subset \Gamma(X^{an},\mathcal{O}_X^{an})$ turns out to be the subset of locally constant functions.)

Now consider those F's in $(X/\mathbb{C})_{inf}$ which are sheaves of $\mathcal{O}_{X/\mathbb{C}}$-modules (or, for that matter, just sheaves of abelian groups). These form an abelian category. Abstract nonsense provides us with enough injectives, so taking the derived functors of the functor $\Gamma(X/\mathbb{C},\ )$ above we get infinitesimal cohomology $H^i(X/\mathbb{C},F)$, Grothendieck proved [4]:

1. If $X/\mathbb{C}$ is smooth, $H^i(X/\mathbb{C},\mathcal{O}_{X/\mathbb{C}}) \cong \mathbb{H}^i(X,\Omega^{\bullet}_{X/\mathbb{C}})$.
2. If $Y/\mathbb{C}$ is smooth and $X \hookrightarrow Y$ is a closed subscheme, then $H^i(X/\mathbb{C},\mathcal{O}_{X/\mathbb{C}}) \cong \mathbb{H}^i(\hat{Y},\hat{\Omega}^{\bullet}_{Y/\mathbb{C}})$, where $\hat{Y}$ is the formal completion of Y along X. (Note that the complex $\hat{\Omega}^{\bullet}_{Y/\mathbb{C}}$ makes sense because $\hat{\Omega}^i_{Y/\mathbb{C}} = \varprojlim \Omega^i_{Y/\mathbb{C}}/I^n_X\Omega^i_{Y/\mathbb{C}}$ and d maps $I^n\Omega^i_{Y/\mathbb{C}}$ into $I^{n-1}\Omega^{i+1}_{Y/\mathbb{C}}$.) Many people* have proved that, even if $X/\mathbb{C}$ is not smooth, the latter is isomorphic to $H^i(X^{an},\mathbb{C})$. Thus, infinitesimal cohomology works well even in the nonsmooth case.
3. If $X' \hookrightarrow X$ is a nilpotent immersion, we get an isomorphism: $H^i(X/\mathbb{C},\mathcal{O}_{X/\mathbb{C}}) \to H^i(X'/\mathbb{C},\mathcal{O}_{X'/\mathbb{C}})$. (This seems plausible from the definition, and it is not difficult to prove
4. The definition of the infinitesimal site can be extended, with an arbitrary base scheme S in place of Spec $\mathbb{C}$, and "base changing" works well. In particular, if X/S is smooth, if $S' \hookrightarrow S$ is a nilpotent thickening, and if $X' \to S'$ is $X\times_S S'$, we get $H^i(X'/S,\mathcal{O}_{X'/S}) \cong H^i(X/S,\mathcal{O}_{X/S}) \cong \mathbb{H}^i(X/S,\Omega^{\bullet}_{X/S})$.

Note that properties 3 and 4 provide us with an instrinsic formulation of the independence of cohomology of a lifting: If X_n is smooth and proper over $S_n = \operatorname{Spec} \mathbb{C}[t]/(t^{n+1})$, we can view X_n as a lifting or deformation of the fiber X_0 over $t = 0$. Then $\mathbb{H}^i(X_n/S_n,\Omega^{\bullet}_{X_n/S_n}) \cong H^i(X_n/S_n,\mathcal{O}_{X_n/S_n}) \cong H^i(X_0/S_n,\mathcal{O}_{X_0/S_n})$.

*c.f. Hartshorne's "On the de Rham Cohomology of Algebraic Varieties" Pub. Math. I.H.E.S. No. 45 (1976) pp. 5-99 for a proof and more references.

These results are expressed "classically" via the Gauss-Manin connection, as explained for instance by Katz and Oda [7].

Now we can try to use this p-adically as follows: S' in 4 is $\underline{\mathrm{Spec}}\ k$, where k has characteristic p, and $S = \underline{\mathrm{Spec}}\ W(k)$. Then $\mathbb{H}^*(X,\Omega^\bullet_{X/S}) \cong \varprojlim \mathbb{H}^*(X_n,\Omega^\bullet_{X_n/S_n})$, where $S_n = \underline{\mathrm{Spec}}\ W_n(k)$ and $X_n = X \times_S S_n$, if X/S is smooth and proper. If the following key properties held, this cohomology group would also be given, purely in terms of X', by $\varprojlim H^*(X/S_n,\mathcal{O}_{X'/S_n})$:

A) $H^*(X_n/W_n,\mathcal{O}_{X_n/W_n}) \cong H^*(X'/W_n,\mathcal{O}_{X'/W_n})$

B) $H^*(X_n/W_n,\mathcal{O}_{X_n/W_n}) \cong \mathbb{H}^*(X_n,\Omega^\bullet_{X_n/W_n})$

Property A) will hold with the above definitions, but Property B) fails. This is because the proof of B) in characteristic zero rests essentially on the Poincaré lemma for the de Rham complex of a formal power series ring — which fails in characteristic $p > 0$. In order to integrate, one needs terms like $t^k/k!$. So, says Grothendieck, let us agree to consider, instead of <u>all</u> nilpotent immersions, only those which are endowed with such "divided powers". This gives us the so-called "crystalline site", which seems to have just the amount of rigidity we need. Before we explain these divided powers, however, we shall review the formalism of differential operators, their linearization, and their relation to descent data.

Perhaps at this point it is desirable to answer the question: "Why bother with the crystalline theory, when ℓ-adic cohomology works so well?" One of the original motivations was to have a meaningful theory of p-torsion. At present, p-torsion remains almost a total mystery – although in principle crystalline cohomology does give reasonable looking p-torsion, very little is known about it. Probably the most important aspect of crystalline cohomology (which distinguishes it from étale cohomology) is its connection to Hodge theory. In particular, it seems to be the best way to attack Katz's conjecture about the relation of the p-adic nature of zeta functions to Hodge numbers (which first emerged in Dwork's work on hypersurfaces [3]). Perhaps even more striking is Mazur's discovery (in the course of his proof of Katz's conjecture) [10] that the Hodge filtration of a suitable variety in characteristic $p > 0$ is <u>determined</u> by the action of Frobenius on $H^i_{cris}(X/W)$. Let me remark that for this result it is crucial to work with cohomology with values in W-modules, not just $W \otimes \mathbb{Q}$-modules.

References for §1

[1] Bloch, S. "K-theory and crystalline cohomology" to appear in Publ. Math. I.H.E.S.

[2] Deligne, P. "La conjecture de Weil I" Publ. Math. I.H.E.S. 43 (1974).

[3] Dwork, B. "On the zeta function of a hypersurface II" Ann. of Math 80 No. 2 (1964) pp. 227-299.

[4] Grothendieck, A. "Crystals and the de Rham cohomology of schemes" in Dix Exposés sur la Cohomologie des schémas North Holland (1968).

[5] ______________. "On the de Rham cohomology of algebraic varieties" Publ. Math. I.H.E.S 29 (1966) pp. 95-103.

[6] Katz, N. "On the differential equations satisfied by period matrices" Publ. Math. I.H.E.S. 35 (1968) pp. 71-106.

[7] Katz, N., and Oda, T. "On the differentiation of DeRham cohomology classes with respect to parameters" J. Math. Kyoto U. 8 (1968), pp. 199-213.

[8] Kleiman, S. "Algebraic cycles and the Weil conjectures" in Dix Exposés sur la Cohomologie des Schémas, North Holland (1968) pp. 359-386.

[9] Lubkin, S. "A p-adic proof of the Weil conjectures" Ann. of Math. 87 (1968) pp. 105-255.

[10] Mazur, B. "Frobenius and the Hodge filtration — estimates" Ann. of Math. 98 (1973) pp. 58-95.

[11] Monsky, P. P-adic Analysis and Zeta Functions Kinokuniya Book Store, Tokyo (1970).

[12] Serre, J. P. "Quelques propriétés des variétés abéliennes en caractéristique p" Am. J. Math. 80 No. 3 (1958) pp. 715-739.

[13] ____________ "Sur la topologie des variétés algébriques en caractéristique p" Symp. Int. de Top. Alg. Univ. Nac. Aut. de Mexico (1958) pp. 24-53.

[14] Zariski, O. Algebraic Surfaces (second supplemented edition) Springer Verlag, (1971).

§2. Calculus and Differential Operators.

In this chapter we develop Grothendieck's way of geometrizing the notions of calculus and differential geometry, and in particular the notion of a locally (or rather infinitesimally) constant sheaf. We begin by reviewing the formalism of differential operators.

If $X \to S$ is a morphism of schemes, and if F and G are $\mathcal{O}_X$-modules, then a differential operator from F to G, relative to S, will be an $f^{-1}(\mathcal{O}_S)$-linear map $h: F \to G$ which is "almost" $\mathcal{O}_X$-linear. In order to make this precise, we begin by brutally linearizing h, i.e., by forming the obvious adjoint map:

$$\bar{h}: \mathcal{O}_X \otimes_{f^{-1}(\mathcal{O}_S)} F \to G \quad .$$

Using the $\mathcal{O}_X$-module structure of F, we can make a natural identification: $\mathcal{O}_X \otimes_{\mathcal{O}_S} F \leftrightarrow P_{X/S} \otimes_{\mathcal{O}_X} F$, where $P_{X/S} = \mathcal{O}_X \otimes_{\mathcal{O}_S} \mathcal{O}_X$. (In order to be kind to the typist, we shall often write $\mathcal{O}_S$ for $f^{-1}(\mathcal{O}_S)$.) With this identification, $\bar{h}$ maps an element of the form $a \otimes b \otimes x$ to $ah(bx)$.

Notice that P has two structures of an $\mathcal{O}_X$-algebra, via the two maps d_0 and d_1: $\mathcal{O}_X \to P_{X/S}$ sending a to $a \otimes 1$ and $1 \otimes a$, respectively. If we form the scheme $P_{X/S} = X \times_S X$, then d_i corresponds to one of the projections $p_i : P_{X/S} \to X$. In writing $P_{X/S} \otimes F$, we use the map d_1 to construct the tensor product and the map d_0 to obtain an $\mathcal{O}_X$-module structure on the result. If X/S is affine, $P_{X/S} \otimes F$ therefore is $p_{0*}p_1^*F$. We shall find it convenient to refer to, and to indicate by writing, the

$\mathcal{O}_X$-module structure of $\mathcal{P}_{X/S}$ from d_0 as the "left" structure and that from d_1 as the "right" structure.

We can summarize our construction by saying that an $\mathcal{O}_S$-linear map $h:\mathcal{F} \to \mathcal{G}$ induces a unique $\mathcal{O}_X$-linear map $\bar{h}:\mathcal{P}_{X/S}\otimes\mathcal{F} \to \mathcal{G}$ such that $\bar{h}\circ d_{1,\mathcal{F}} = h$, where $d_{1,\mathcal{F}}:\mathcal{F} \to \mathcal{P}_{X/S}\otimes\mathcal{F}$ is the map $d_1 \otimes id_{\mathcal{F}}$.

It is easy to tell from $\bar{h}$ whether or not h is linear: Let $\mathcal{I} \subseteq \mathcal{P}_{X/S}$ be the kernel of the map induced by multiplication $\mathcal{P}_{X/S} \to \mathcal{O}_X$, i.e., the ideal of the diagonal Δ in $X\times_S X$. Note that $\mathcal{I}$ is generated as $\mathcal{O}_X$-module by elements of the form $1\otimes b - b\otimes 1$, with $b \in \mathcal{O}_X$. Indeed, if $\alpha = \sum_i a_i\otimes b_i \in I$, then $\sum_i a_i b_i = 0$, and hence: $\alpha=\sum_i(a_i\otimes 1)(1\otimes b_i - b_i\otimes 1)$.

Now h is $\mathcal{O}_X$-linear iff $h(bx) = bh(x)$ for all $b \in \mathcal{O}_X$, i.e. iff $\bar{h}(1\otimes b\otimes x) = \bar{h}(b\otimes 1\otimes x)$, i.e., iff $\bar{h}$ annihilates $\mathcal{I}(\mathcal{P}\otimes\mathcal{F})$. This makes, I hope, the following definition of differential operator a reasonable notion of "almost" linear:

2.1 Definition. An $f^{-1}(\mathcal{O}_S)$-linear map $h: \mathcal{F} \to \mathcal{G}$ is a "differential operator of order $\leq n$" iff $\bar{h}$ annihilates $\mathcal{I}^{n+1}\otimes\mathcal{F}$; equivalently, iff h factors:

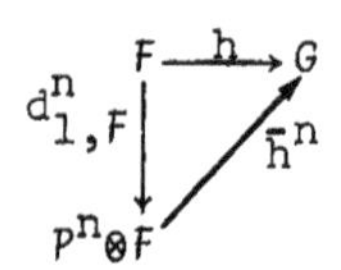

where $\bar{h}^n$ is $\mathcal{O}_X$-linear, $\mathcal{P}^n = \mathcal{P}/\mathcal{I}^{n+1}$, and $d^n_{1,\mathcal{F}}$ is induced by $x \mapsto 1\otimes 1\otimes x$. In other words, $d^n_{1,\mathcal{F}}$ is the universal differential operator of order $\leq n$.

Before proceeding, let us give an explicit description of $P_{X/S}$ if X/S is smooth, in local coordinates. Notice first that $P_{X/S}$ is generated as a left $\mathcal{O}_X$-module by all elements of the form $d_1(b)$: $b \in \mathcal{O}_X$, since any $\alpha \in P_{X/S}$ can be written as a sum $\alpha = \Sigma a_i \otimes b_i = a_i d_1(b_i)$. More generally, if E is any $\mathcal{O}_X$-module, $P \otimes E$ is generated, as an $\mathcal{O}_X$-module, by $\{d_{1,E}(x) : x \in E\}$. Of course, analogous statements hold with P^m in place of P.

2.2 <u>Proposition.</u> Suppose X/S is smooth and $x_1 \ldots x_n$ are local coordinates, i.e., sections of $\mathcal{O}_X$ defining an étale map $X \to A^n_S$. Let $\xi_i = 1 \otimes x_i - x_i \otimes 1$. Then $P^m_{X/S}$ is the free $\mathcal{O}_X$-module with basis the image of $\{\xi_1^{\alpha_1} \ldots \xi_n^{\alpha_n} : \Sigma \alpha_i \leq m\}$. (This holds with both the left and right $\mathcal{O}_X$-structures on P^m; symmetry allows us to consider only the left structure.)

<u>Proof.</u> Since $\Delta : X \hookrightarrow X \times_S X$ is a locally closed immersion of smooth S-schemes, it is a regular immersion. Now the hypotheses imply that $dx_1 \ldots dx_n \in \Omega^1_{X/S} = I/I^2$ are a basis of I/I^2, and hence $Gr^{\cdot}_I(P) \cong \mathcal{O}_X[dx_1 \ldots dx_n]$. Since dx_i is the class of ξ_i in I/I^2, we see that $Gr^j_I(P) = I^j/I^{j+1}$ is free with basis the images of the monomials of degree j in the ξ_i's. Thus the exact sequences: $0 \to I^j/I^{j+1} \to P^j \to P^{j-1} \to 0$ and induction make the result clear. □

In order to understand composition of differential operators, let us go back to P. Suppose $f : F \to G$ and $g : G \to H$ are $f^{-1}(\mathcal{O}_S)$-linear; how can we describe $\overline{g \circ f} : P \otimes F \to H$ in terms of $\bar{f}$ and $\bar{g}$? Clearly $\overline{g \circ f} = \bar{g} \circ (id_{\mathcal{O}_X} \otimes_{\mathcal{O}_S} f)$, so the problem is how to recover $id_{\mathcal{O}_X}\otimes_{\mathcal{O}_S} f\colon \mathcal{O}_X\otimes_{\mathcal{O}_S}\mathcal{F} \to \mathcal{O}_X\otimes_{\mathcal{O}_S}\mathcal{G}$ from $\bar{f}\colon \mathcal{O}_X\otimes_{\mathcal{O}_S}\mathcal{F} \cong \mathcal{P}\otimes_{\mathcal{O}_X}\mathcal{F} \to \mathcal{G}$. To do this, consider the map $\delta\colon\mathcal{P}\to\mathcal{P}\otimes_{\mathcal{O}_X}\mathcal{P}$ given by $a\otimes b \mapsto a\otimes_{\mathcal{O}_S}1\otimes_{\mathcal{O}_X}1\otimes_{\mathcal{O}_S}b = a\otimes1\otimes b$. If we identify $\mathcal{P}\otimes_{\mathcal{O}_X}\mathcal{P}$ with $\mathcal{O}_X\otimes_{\mathcal{O}_S}\mathcal{O}_X\otimes_{\mathcal{O}_S}\mathcal{O}_X$, δ corresponds to the geometric map $X\times X\times X \to X\times X\colon (x,y,z)\mapsto(x,z)$. Clearly δ is a ring homomorphism and is $\mathcal{O}_X$-linear if $\mathcal{P}$ and $\mathcal{P}\otimes_{\mathcal{O}_X}\mathcal{P}$ are (simultaneously) viewed as $\mathcal{O}_X$-modules by the extreme left or right. Thus $\delta\otimes id_{\mathcal{F}}\colon \mathcal{P}\otimes_{\mathcal{O}_X}\mathcal{F} \to \mathcal{P}\otimes_{\mathcal{O}_X}\mathcal{P}\otimes_{\mathcal{O}_X}\mathcal{F}$ makes sense and is $\mathcal{O}_X$-linear (using the left structure), hence so is

$$\bar{\delta}(f) \underset{\text{def.}}{=} (id_{\mathcal{P}}\otimes_{\mathcal{O}_X}\bar{f})\circ(\delta\otimes_{\mathcal{O}_X}id_{\mathcal{F}})\colon \mathcal{P}\otimes_{\mathcal{O}_X}\mathcal{F} \to \mathcal{P}\otimes_{\mathcal{O}_X}\mathcal{G} .$$

2.3 Lemma. With the above notation and the identification $\mathcal{P}\otimes_{\mathcal{O}_X} \cong \mathcal{O}_X\otimes_{\mathcal{O}_S}$

1) $\bar{\delta}(f) = id_{\mathcal{O}_X}\otimes_{\mathcal{O}_S} f : \mathcal{P}\otimes_{\mathcal{O}_X}\mathcal{F} \to \mathcal{P}\otimes_{\mathcal{O}_X}\mathcal{G}$

2) $(\overline{g\circ f}) = \bar{g}\circ\bar{\delta}(f) : \mathcal{P}\otimes_{\mathcal{O}_X}\mathcal{F} \to \mathcal{H}$.

3) The map δ induces maps:

$$\begin{array}{ccc} \mathcal{P} & \xrightarrow{\ \delta\ } & \mathcal{P}\otimes\mathcal{P} \\ \downarrow & & \downarrow \\ \mathcal{P}^{n+m} & \xrightarrow{\ \delta^{n,m}\ } & \mathcal{P}^{n}\otimes\mathcal{P}^{m} \end{array}$$

Proof. To check 1), it suffices to consider elements of the form $a\otimes_{\mathcal{O}_S}1\otimes_{\mathcal{O}_X}x$ in $\mathcal{P}\otimes_{\mathcal{O}_X}\mathcal{F}$, which go by $\delta\otimes_{\mathcal{O}_X}id_{\mathcal{F}}$ to

$a\otimes_{\mathcal{O}_S} 1\otimes_{\mathcal{O}_X} 1\otimes_{\mathcal{O}_S} 1\otimes_{\mathcal{O}_X} x$ and then by $id_P\otimes\bar{f}$ to $a\otimes_{\mathcal{O}_S} 1\otimes_{\mathcal{O}_X} f(x)$. As we have observed, 2) follows; it can also be deduced from the universal mapping property of $P\otimes_{\mathcal{O}_X} F$. For 3), we must check that the clockwise composition annihilates I^{n+m+1}. Recall that the ideal I is generated by elements of the form $\xi = 1\otimes x - x\otimes 1$. Then $1\otimes 1\otimes x\otimes 1 = 1\otimes x\otimes 1\otimes 1$, so $\delta(\xi) = 1\otimes 1\otimes 1\otimes x - x\otimes 1\otimes 1\otimes 1 =$ $(1\otimes 1)\otimes(1\otimes x - x\otimes 1) + (1\otimes x - x\otimes 1)\otimes 1\otimes 1$, i.e.

(2.3.4) $$\delta(\xi) = \mathbb{1}\otimes\xi + \xi\otimes\mathbb{1}\ , \quad \text{for}\quad \xi = 1\otimes x - x\otimes 1,\ \mathbb{1} = 1\otimes 1\ .$$

Now I^{n+m+1} is generated by products $\Pi = \xi_1\cdots\xi_{n+m+1}$, and since δ is a ring homomorphism

$$\delta(\Pi) = \prod_{i=1}^{m+n+1} (\mathbb{1}\otimes\xi_i + \xi_i\otimes\mathbb{1})\ .$$

If we expand this as a sum, it is clear that each term must have at least $m+1$ ξ_i's to the right of the $\otimes$, or $n+1$ to the left. In either case, the image of the term in $P^n\otimes P^m$ is zero.

2.4 Corollary. If $f:F \to G$ and $g:G \to H$ are differential operators of orders $\leq n$ and m respectively, then the composition $g\circ f:F \to H$ is a differential operator of order $\leq m+n$, and we have a commutative diagram:

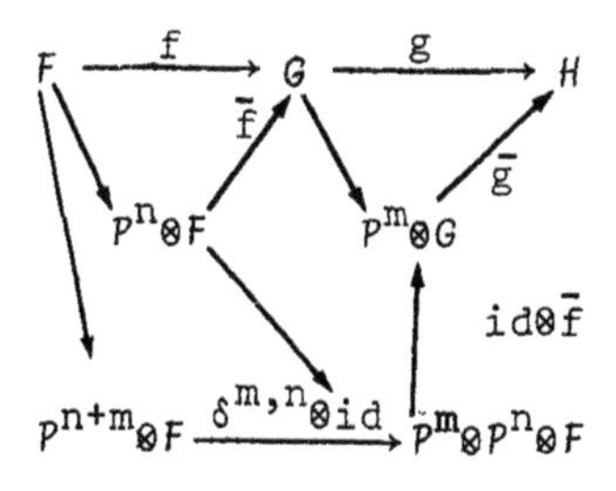

2.3 Remark. The nontrivial commutativity in the diagram is the lower left triangular region; its commutativity gives another explanation of 1) and 2) in the above lemma. It is obtained by tensoring the square below with F .

If E and F are $\mathcal{O}_X$-modules, we let $\mathcal{Diff}^n_{X/S}(E,F)$ denote the sheaf of germs of differential operations of order $\leq n$ from E to F, which can be canonically identified with $\mathcal{Hom}_{\mathcal{O}_X}(P^n \otimes E, F)$. Notice that the $\mathcal{O}_X$-module structure is compatible with the usual one on F ; thus if $D: E \to F$ is a differential operator and α is a section of $\mathcal{O}_X$, αD is the composition "α"$\circ D$, regarding multiplication by α as a differential operator "α" of order 0. We let $\mathcal{Diff}_{X/S}(E,F) = \varinjlim_n \mathcal{Diff}^n_{X/S}(E,F)$. In fancy language, the functor $\mathcal{Diff}_{X/S}(E, \ \ )$ is represented by the pro-object "lim"$\{P^n \otimes E : n \in \mathbb{N}\}$.

When X/S is smooth, we can locally give an explicit description of differential operators and their composition:

2.6 Proposition. Suppose X/S is smooth, $x_1, \ldots x_n$ are local coordinates, and ξ_i is the image of $1 \otimes x_i - x_i \otimes 1$ in $P^m_{X/S}$. If $q = (q_1, \ldots q_n)$ is a multi-index, with $q_i \in \mathbb{N}$, then ξ^q denotes $\prod_{i=1}^n \xi_i^{q_i}$, $q! = \prod_{i=1}^n (q_i!)$, and $|q| = \sum_{i=1}^n q_i$, so that $\{\xi^q : |q| \leq m\}$ is a basis for P^m. Let $\bar{D}_q$ be the dual basis for $Hom(P^m, \mathcal{O}_X)$, and let $D_q : \mathcal{O}_X \to \mathcal{O}_X$ be the corresponding differential operator. Then $\{D_q : |q| \leq m\}$ is a basis for $\mathcal{D}i\!f\!f^m_{X/S}(\mathcal{O}_X, \mathcal{O}_X)$, viewed as an $\mathcal{O}_X$-module through the second $\mathcal{O}_X$, and composition is given by:

$$D_q \circ D_{q'} = \frac{(q+q')!}{(q!)(q')!} D_{q+q'} \underset{\text{def}}{=} \binom{q+q'}{q} D_{q+q'} \quad .$$

Proof. If $\bar{D} : P^m \to \mathcal{O}_X$ is $\mathcal{O}_X$-linear and $D : \mathcal{O}_X \to \mathcal{O}_X = \bar{D} \circ d_1^m$ is the corresponding operator, then for any $a \in \mathcal{O}_X$, the operator corresponding to $a \cdot \bar{D}$ is (multiplication by a)$\circ D$. This shows that $\{D_q\}$ forms a basis for $\mathcal{D}i\!f\!f^m(\mathcal{O}_X, \mathcal{O}_X)$ viewed as an $\mathcal{O}_X$-module as we have described. To verify the composition formula, we must show that the map $\overline{D_q \circ D_{q'}} : P^{m+m'} \to \mathcal{O}_X$ takes ξ^p to $\binom{q+q'}{q} \delta_{q+q',p}$, where $\delta_{a,b}$ is Kronecker's δ-function. From the definition of composition, $\overline{D_q \circ D_{q'}}$ is given by $\bar{D}_q \circ (id_{P^m} \otimes \bar{D}_{q'}) \circ \delta$, where $\delta : P^{m+m'} \to P^m \otimes P^{m'}$ is induced by our old δ. Recalling that $\delta(\xi_i) = \mathbb{1} \otimes \xi_i + \xi_i \otimes \mathbb{1}$ and is a homomorphism, we compute

$$\delta(\xi^p) = (\mathbb{1} \otimes \xi + \xi \otimes \mathbb{1})^p = \sum_{i+j=p} \binom{p}{i} (\mathbb{1} \otimes \xi)^i (\xi \otimes \mathbb{1})^j = \sum_{i+j=p} \binom{p}{i} \xi^j \otimes \xi^i \quad .$$

If we apply $id \otimes D_{q'}$, we get zero unless $i = q'$, and so we are left only with $\binom{p}{q'} \xi^{p-q'}$. Applying $\bar{D}_q$, we get zero unless $p - q' = q$, i.e., $p = q+q'$, and in this case we get $\binom{q+q'}{q'}$.

2.7 Remark. The formula implies that the D_q's all commute (although of course they don't commute with, for instance, all operators of order zero). Moreover, over $\mathbb{Q}$, we can write $D_q = \frac{1}{q!} D^q$ where $D^q = \prod_{i=1}^{n} D_i^{q_i}$ and $D_i = D_{(0,0\ldots,1,\ldots 0)}$. Thus $\mathcal{D}iff^1(\mathcal{O}_X,\mathcal{O}_X)$ generates $\mathcal{D}iff(\mathcal{O}_X,\mathcal{O}_X)$ – a fact which fails in characteristic p, or over $\mathbb{Z}$.

We can now try to use differential operators to develop a suitable notion of locally constant sheaves.[1] We begin with the familiar:

2.8 Definition. A "connection" on an $\mathcal{O}_X$-module E is an additive map: $\nabla: E \to E\otimes\Omega^1_{X/S}$ such that $\nabla(ax) = a\nabla x + x\otimes da$ if x is a section of E and a is a section of $\mathcal{O}_X$.

For example, the exterior derivative $d: \mathcal{O}_X \to \Omega^1_{X/S}$ is a connection, called the "constant" one. It is easy to see that the set of all connections on $\mathcal{O}_X$ is in one-one correspondence with $\Gamma(X,\Omega^1_{X/S})$, via $\nabla \leftrightarrow \nabla(1)$.

2.9 Proposition. A connection on E is equivalent to a P^1-linear isomorphism $\varepsilon: P^1\otimes E \to E\otimes P^1$ which, modulo the kernel $\Omega^1_{X/S}$ of $P^1 \to \mathcal{O}_X$, reduces to the identity endomorphism of E.

Proof. Starting from ε above, let $\theta = \varepsilon\circ d_{1_E}: E \to E\otimes P^1$. Since $d_{1,E}$ is linear for the right module structure of P^1, so is θ, i.e. $\theta(ax) = (1\otimes a)\theta(x)$. Now let $\nabla(x) = \theta(x) - x\otimes 1$; note that $\nabla(x) \in E\otimes\Omega^1$ because ε reduces to the identity. Now compute:

[1] For an explanation of the relationship between locally constant sheaves and connections in the classical case, we refer the reader to Deligne's Equations Différentielles... LNM 163, Springer.

$$\nabla(ax) = \theta(ax) - ax\otimes 1 = (1\otimes a)\,\theta(x) - (a\otimes 1)(x\otimes 1)$$

$$= (1\otimes a)\,\theta(x) - (1\otimes a)(x\otimes 1) + (1\otimes a)(x\otimes 1) - (a\otimes 1)(x\otimes 1)$$

$$= (1\otimes a)\,\nabla(x) + (da)(x\otimes 1) = a\nabla(x) + x\otimes da \quad .$$

(Recall that $(1\otimes a)\,\omega = (a\otimes 1)\omega = a\omega$ for $\omega \in \Omega^1_{X/S}$.)

Conversely, given ∇ , let $\theta(x) = \nabla(x) + x\otimes 1$; then reversing the previous calculation shows that $\theta : E \to E\otimes P^1$ is linear for the right $\mathcal{O}_X$-structure on $E\otimes P^1$. Extension of scalars gives us the P^1-linear map $\varepsilon : P^1\otimes E \to E\otimes P^1$, which clearly reduces to the identity mod Ω^1. To see that ε is an isomorphism, consider the involution $\tau : P^1 \to P^1$ induced by switching the factors; $a\otimes b \mapsto b\otimes a$. We have a τ-linear map $\sigma : E\otimes P^1 \to P^1\otimes E : x\otimes\xi \longrightarrow \tau(\xi)\otimes x$; of course σ is bijective and also reduces to the identity mod Ω^1. The endomorphism $(\sigma\circ\varepsilon)$ of $P^1\otimes E$ is τ-linear, so $(\sigma\circ\varepsilon)^2$ is linear, moreover it reduces to the identity modulo the square zero ideal Ω^1. It follows that $(\sigma\circ\varepsilon)^2$ is an isomorphism, hence $(\sigma\circ\varepsilon)$ is bijective, and hence so is ε . □

Let us now try to motivate Grothendieck's description of a connection in terms of a suitable site. Recall that descent data for a sheaf E on X relative to $f : X \to S$ means an isomorphism $p_2^*E \to p_1^*E$ on $P_{X/S} = X\times_S X$, satisfying certain transitivity conditions. If $E \cong f^*K$ for some sheaf of $\mathcal{O}_S$-modules K, E has natural descent data, which we shall say is "effective" or "constant". (A formula for the data in this case is given on page 2.17 .) Now for any T lying between X and S, we can descend E to a sheaf E_T on T (namely the pullback of K), and these sheaves will be compatible in the obvious sense.

By (2.9), a connection is nothing more than <u>first</u> <u>order</u> descent data. (We are temporarily postponing a discussion of integrability and the cocycle condition.) In particular, if T lies between X and S <u>and</u> is such that $X \to T$ is a closed immersion defined by a square zero ideal, we can hope to find a natural sheaf E_T on T. First suppose there were a retraction $g_1 : T \to X$; we would then be happy to take $E_T = g_1^*(E)$, were it independent of g_1. In fact, if $g_2 : T \to X$ is another retraction, $g = (g_1, g_2) : T \to X_1 \times X_2$ factors through the first infinitesimal neighborhood $P^1_{X/S} = \mathcal{S}pec_X \mathcal{P}^1_{X/S}$ of the diagonal in $P_{X/S}$, since g_1 and g_2 agree modulo a square zero ideal. Now the data of a connection is an isomorphism $\varepsilon : p_2^*E|_{P^1} \to p_1^*E|_{P^1}$; pulling it back via g gives an isomorphism $g^*(\varepsilon) : g_2^*E \to g_1^*E$. Hence up to (so far non-canonical) isomorphism, E_T is indeed independent of the retraction. One deduces that if T' is another "first order thickening" of X admitting a retraction, and if $u : T' \to T$ is compatible with the retractions, there is an isomorphism $\rho_u : u^*E_T \to E_{T'}$. A suitable cocycle condition would give us compatibility of these isomorphism with composition, and also would allow us to construct the sheaves E_T under the (much more natural) condition that the retractions $T \to X$ existed only locally, (since we could then glue the local constructions).

The compatibility which we shall need is the following: Suppose that T'' is another thickening of X, and that $v : T'' \to T'$ is another morphism. Then we require that the following diagram should commute:

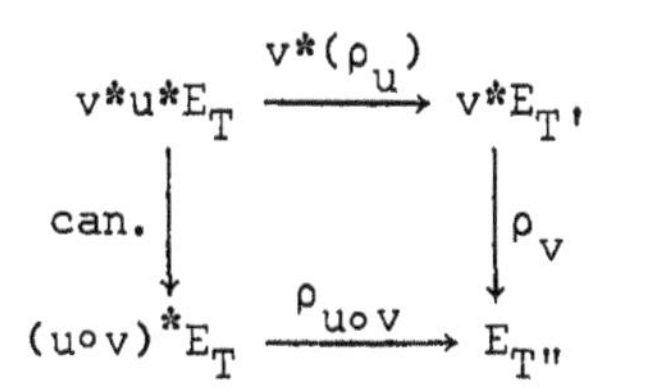

We shall make all this precise later. For now, I hope the reader is convinced that the notion of an (integrable) connection is equivalent to first order decent data, which should in turn be equivalent to the data of a sheaf on a site made up of the first order thickenings of (open subsets of) X. This generalizes to higher orders as well:

2.10 <u>Definition.</u> A "stratification" on E is a collection of isomorphisms ε_n: $P^n_{X/S} \otimes E \to E \otimes P^n_{X/S}$ such that:

1) ε_n is $P^n_{X/S}$-linear.
2) ε_n and ε_m are compatible, via the "restriction"
$$P^n_{X/S} \to P^m_{X/S} \quad , \quad \text{for} \quad m \leq n \; .$$
3) ε_0 is the identity map.
4) The cocycle condition holds: If $P^n_{X/S}(2)$ is the nth infinitesimal neighborhood of X in $X \times_S X \times_S X$, and p_{ij}: $P^n_{X/S}(2) \to P^n_{X/S}$ is projection via the coordinates i and j, then for all n:

$$p^*_{12}(\varepsilon_n) \circ p^*_{23}(\varepsilon_n) = p^*_{13}(\varepsilon_n) \; .$$

Let me try to explain the cocycle condition. We shall let $\pi_i : P^n_{X/S}(2) \to X$ and $p_i : P^n_{X/S} \to X$ stand for the ith projection. Note that:

$$\pi_1 = p_1 \circ p_{13} = p_1 \circ p_{12}$$

$$\pi_2 = p_2 \circ p_{12} = p_1 \circ p_{23}$$

$$\pi_3 = p_2 \circ p_{23} = p_2 \circ p_{13}$$

Thus, the cocycle condition says that the following diagram commutes for all n.

4a) 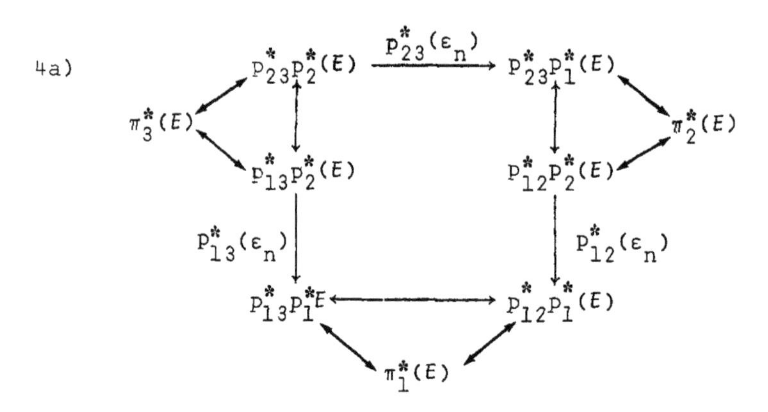

In other words, it says that if we use the stratification ε_n to construct isomorphisms $\pi_3^*(E) \to \pi_2^*(E)$ and $\pi_2^*(E) \to \pi_1^*(E)$ then the composition of these is the isomorphism $\pi_3^*(E) \to \pi_1^*(E)$ provided by ε_n. We leave it as an exercise for the reader to deduce the compatibility of the ρ_u's described above.

We can describe this algebraically: Identifying $P_{X/S}(2) = \mathcal{O}_X \otimes_{\mathcal{O}_S} \mathcal{O}_X \otimes_{\mathcal{O}_S} \mathcal{O}_X$ with $P \otimes_{\mathcal{O}_X} P$, the map p_{13}^* becomes the map $\delta: P \to P \otimes P$. $P(2)$ has three $\mathcal{O}_X$-module structures, and the $\pi_i^*(E)$ correspond to $E \otimes P \otimes P$, $P \otimes E \otimes P$, and $P \otimes P \otimes E$ for $i=1,2,3$, respectively. The cocycle condition can then also be expressed:

4b) The following diagram commutes, for all m and n:

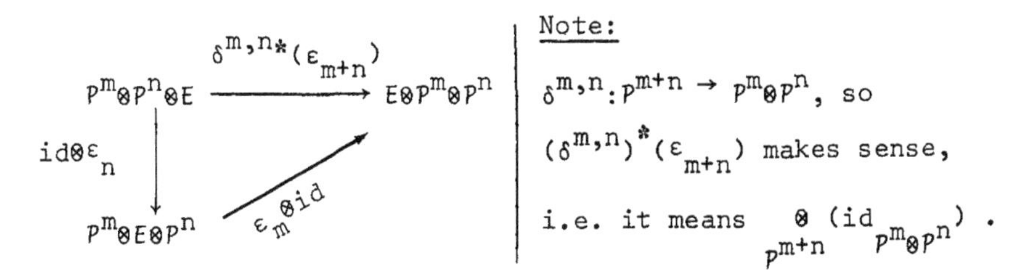

To motivate some of the other ways of giving a stratification on the $\mathcal{O}_X$-module E return to the case of effective descent data for $f:X \to S$. If $E \cong f^*(K)$, for some $\mathcal{O}_S$-module K, then for any $\mathcal{O}_X$-module F , $E\otimes_{\mathcal{O}_X} F \cong f^{-1}(K)\otimes_{f^{-1}(\mathcal{O}_S)} F$. Thus we get a natural transformation, for any two $\mathcal{O}_X$-modules F and G ,

$\mathcal{H}om_{\mathcal{O}_S}[F,G] \to \mathcal{H}om_{\mathcal{O}_S}[E\otimes_{\mathcal{O}_X} F, E\otimes_{\mathcal{O}_X} G]$, by $h \mapsto id_{f^{-1}(K)}\otimes_{\mathcal{O}_S} h$.

We shall see that a stratification on E allows us to do this if h is a differential operator.

2.11 <u>Proposition.</u> Suppose X/S is smooth and E is an $\mathcal{O}_X$-module. The following data are equivalent:

1) A stratification on E (i.e., the maps ε_n above).

2) A collection of compatible, right $\mathcal{O}_X$-linear maps $\theta_n: E \to E\otimes P^n_{X/S}$, with $\theta_0 = id_E$, such that the following diagram commutes (cocycle condition):

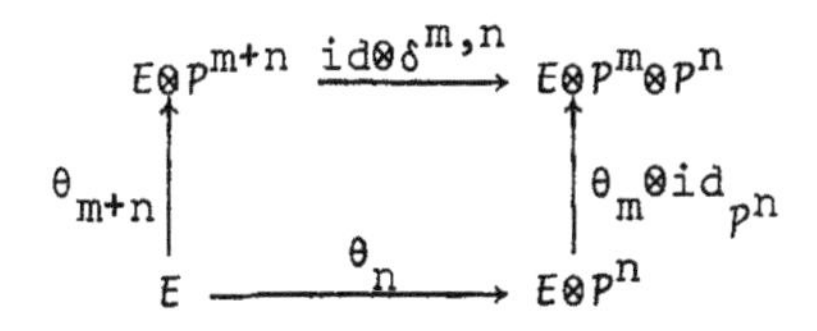

3) An $\mathcal{O}_X$-linear ring homomorphism:

$$\mathcal{D}iff_{X/S}(\mathcal{O}_X,\mathcal{O}_X) \to \mathcal{D}iff_{X/S}(E,E) .$$

3 bis) A collection of $\mathcal{O}_X$-linear maps, for any two $\mathcal{O}_X$-modules F and G: $\nabla:\mathcal{D}iff_{X/S}(F,G) \to \mathcal{D}iff_{X/S}(E\otimes F,E\otimes G)$, compatible with composition, and taking the identity of $\mathcal{D}iff(F,F)$ to the identity.

4) A compatible family of sheaves E_T for each nilpotent thickening $U \longrightarrow T$ of an open set in X, with transitive isomorphisms $u^*E_T \to E_{T'}$, for any $u:(U',T') \to (U\ ,T\ )$.

2.12 Remark. Notice that in 4) the "transition maps" are isomorphisms, so $\{E_T\}$ is not just any object of the infinitesimal topos of X/S. In the terminology of Grothendieck's article in Dix Exposés, $\{E_T\}$ is "special", these days we might call $\{E_T\}$ a crystal in the infinitesimal topos.

Sketch of the Equivalence. The equivalence of (1) and (4) has already been sketched. Given the data of the ε's, one gets the θ's by $\theta_n = \varepsilon\circ(d_1\otimes id)$: $E \to P^n\otimes E \to E\otimes P^n$, we leave to the reader the fact that the cocycle condition translates as claimed.

Given the θ_n's, we get the data of (3bis) (and hence of (3)) as follows: If $h:F \to G$ is a differential operator of order $\leq n$, and $\bar{h}:P^n\otimes F \to G$ is its $\mathcal{O}_X$-linearization, we let $\nabla(h) = (id_E\otimes\bar{h})\circ(\theta_n\otimes id_F)$: $E\otimes F \to E\otimes P^n\otimes F \to E\otimes G$, which is a differential operator of order $\leq n$. The fact that $\nabla(id_F) = id_{E\otimes F}$ is

equivalent to the fact that $\theta_0 = \mathrm{id}_E$. Indeed, if $\mu: P \to \mathcal{O}_X \cong P^0$ is the projection, then $\overline{\mathrm{id}}_F: P^n \otimes F \to F$ is given by $\xi \otimes x \mapsto \mu(\xi) \otimes x$, and $\mathrm{id}_E \otimes \overline{\mathrm{id}}_F$: $E \otimes P^n \otimes F \to E \otimes F$ is $e \otimes \xi \otimes x \mapsto \mu(\xi) e \otimes x = (\mu_n \otimes \mathrm{id}_F)(e \otimes \xi \otimes x)$, where $\mu_n: E \otimes P^n \to E \otimes P^0 \cong E$ is the projection. Thus
$\nabla(\mathrm{id}_F) = (\mu_n \otimes \mathrm{id}_F) \circ (\theta_n \otimes \mathrm{id}_F) = (\mu_n \circ \theta_n) \otimes \mathrm{id}_F = \theta_0 \otimes \mathrm{id}_F$.

Finally, the cocycle condition is equivalent to the fact that ∇ preserves compositions. We shall verify one direction, and with the reader's permission, shall suppress the subscripts First of all, if $f: F \to G$ is a differential operator, I claim that the following diagram commutes:

$$\begin{array}{ccc} E \otimes G & \xrightarrow{\theta \otimes \mathrm{id}_G} & E \otimes P \otimes G \\ \nabla(f) \Big\uparrow & & \Big\uparrow \mathrm{id}_E \otimes \bar{\delta}(f) \\ E \otimes F & \xrightarrow{\theta \otimes \mathrm{id}_F} & E \otimes P \otimes F \end{array}$$

c.f. Lemma (2.3) (and the preceeding discussion) for the definition of $\bar{\delta}(f)$.

This is a straightforward consequence of the definitions of $\nabla(f)$ and $\bar{\delta}(f)$ and the cocycle condition. Adding the triangle on the right below, which comes from the discussion of composition of differential operators, Lemma (2.3), we get if $g: G \to H$:

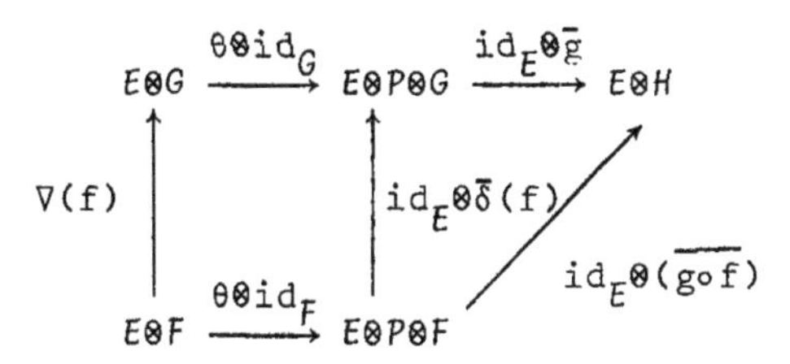

Going around the top is $\nabla(g) \circ \nabla(f)$, around the bottom is $\nabla(g \circ f)$.

Given the data of 3bis) we clearly have the data of 3), which we now show gives 1). View the map

$\nabla_n : \mathcal{D}iff^n(\mathcal{O}_X,\mathcal{O}_X) \to \mathcal{D}iff^n(E,E)$ as a map $\mathcal{H}om_{\mathcal{O}_X}(\mathcal{P}^n,\mathcal{O}_X) \to \mathcal{H}om_{\mathcal{O}_X}(\mathcal{P}^n \otimes E, E)$, where $\mathcal{P}^n$ and $\mathcal{P}^n \otimes E$ have the $\mathcal{O}_X$-structure from the left. The map is $\mathcal{O}_X$-linear, and I claim it is automatically $\mathcal{P}$-linear. To see this, first observe that if $\partial \in \mathcal{D}iff^n(F,G)$ and $x \in \mathcal{O}_X$, then $d_1(x)\partial = \partial \circ \mu_x$, where $\mu_x \in \mathcal{D}iff^0(F,F)$ is multiplication by x, while $d_0(x)\partial$ (which we write $x\partial$) is $\mu_x \circ \partial$, with $\mu_x \in \mathcal{D}iff^0(G,G)$. Check this by recalling that the $\mathcal{P}$-module structure of $\mathcal{D}iff^n(F,G) \cong \mathcal{H}om_{\mathcal{O}_X}(\mathcal{P}^n \otimes F, G)$ is defined by $\xi\partial = \partial \circ (\mu_{\bar{\xi}} \otimes id_F)$ where $\mu_{\bar{\xi}} : \mathcal{P}^n \to \mathcal{P}^n$ is multiplication by the class of ξ in $\mathcal{P}^n$. Thus $d_1(x)\partial = \partial \circ (\mu_{\overline{1 \otimes x}} \otimes id_F) = \partial \circ \mu_x$, while $d_0(x)\partial = \partial \circ \mu_{\overline{x \otimes 1}} \otimes id_F = \mu_{\bar{x}} \circ \partial$ because ∂ is $\mathcal{O}_X$-linear. Now since the maps $\{\nabla_n\}$ preserve composition, $\nabla_n(d_1(x)\partial) = \nabla_n(\partial \circ \mu_x) = \nabla_n(\partial) \circ \nabla_0(\mu_x) = \nabla_n(\partial) \circ \mu_x = d_1(x)\nabla_n(x)$. Since ∇_n is $\mathcal{O}_X$-linear, and since $\{d_1(x) : x \in \mathcal{O}_X\}$ generates $\mathcal{P}$, it follows that $\nabla_n(\xi\partial) = \xi\nabla_n(\partial)$ for all $\xi \in \mathcal{P}$.

Now apply Hom($, E)$ to the map ∇_n to get the map ∇^* below. The map η is evaluation. Finally, since X/S is smooth, $\mathcal{P}^n_{X/S}$ is locally free, so the map β is an isomorphism. Thus, the diagram defines ε_n. We let the reader verify that ε_n has the desired properties.

$$\begin{array}{ccc} \mathcal{P}^n \otimes E & \xrightarrow{\eta} & \mathcal{H}om_{\mathcal{O}_X}(\mathcal{H}om_{\mathcal{O}_X}(\mathcal{P}^n \otimes E, E), E) \\ \downarrow \varepsilon_n & & \downarrow \nabla^* \\ E \otimes \mathcal{P}^n & \xleftrightarrow{\beta} & \mathcal{H}om_{\mathcal{O}_X}(\mathcal{H}om(\mathcal{P}^n, \mathcal{O}_X), E) \end{array}$$

Note: $\mathcal{H}om(\mathcal{P}^n,\mathcal{O}_X)$ is computed using the left $\mathcal{O}_X$-module structure, so that β induces an isomorphism with $E \otimes \mathcal{P}^n$ (rather than $\mathcal{P}^n \otimes E$), as shown. □

2.13 <u>Remark.</u> It follows automatically that any $\mathcal{O}_X$-linear ring homomorphism $\nabla: \mathcal{D}iff(\mathcal{O}_X,\mathcal{O}_X) \to \mathcal{E}nd_{\mathcal{O}_S}(E,E)$ takes differential operators of order $\leq n$ to differential operators of order $\leq n$, because an endomorphism h of E belongs to $\mathcal{D}iff^n(E,E)$ iff $\mathrm{ad}_a(h)$ belongs to $\mathcal{D}iff^{n-1}(E,E)$ for all $a \in \mathcal{O}_X$ [EGA IV, 16.8.8].

What are the maps ∇, θ, ε, in the case of the constant connection on $\mathcal{O}_X$? Well, $\nabla: \mathcal{O}_X \to \Omega^1_{X/S} \subseteq P^1_{X/S}$ is just the exterior derivative d, which is induced by $x \mapsto 1\otimes x - x\otimes 1$. Recalling that $\theta(x) = \nabla(x) + x\otimes 1$, we see that $\theta(x)$ is the reduction of $1\otimes x$, and since $\varepsilon: P^1 \to P^1$ is the linearization of θ, $\varepsilon(a\otimes x) = a\varepsilon(1\otimes x)$ $= a\varepsilon(x) = a\otimes x$ — i.e. ε is just the identity map.

This works for any relatively constant connection on a sheaf of the form $f^*(K)$, where K is a sheaf on S. Slightly more generally, let us suppose that E is a sheaf of $f^{-1}(\mathcal{O}_S)$-modules on X. Then we can put a constant connection on $L = \mathcal{O}_X \otimes_{\mathcal{O}_S} E$ by taking the reductions mod n of the following maps:

Begin with the map $\tau: \mathcal{O}_X \otimes_{\mathcal{O}_S} E \to E \otimes_{\mathcal{O}_S} \mathcal{O}_X$ sending $c\otimes x$ to $x\otimes c$. There is a natural identification: $\mathcal{O}_X \otimes_{\mathcal{O}_S} E \otimes_{\mathcal{O}_S} \mathcal{O}_X \cong (\mathcal{O}_X \otimes_{\mathcal{O}_S} E) \otimes_{\mathcal{O}_X} \mathcal{O}_X \otimes_{\mathcal{O}_S} \mathcal{O}_X$ which takes $b\otimes x\otimes c$ to $b\otimes x\otimes 1\otimes c = 1\otimes x\otimes b\otimes c$. If we compose τ with the "inclusion" $E \otimes_{\mathcal{O}_S} \mathcal{O}_X \to \mathcal{O}_X \otimes_{\mathcal{O}_S} E \otimes_{\mathcal{O}_S} \mathcal{O}_X$ and make these identifications, we obtain a map $\varepsilon: L \to L \otimes_{\mathcal{O}_X} P_{X/S}$, which maps $c\otimes x$ to $1\otimes x\otimes 1\otimes c$. The induced $\varepsilon: P_{X/S} \otimes_{\mathcal{O}_X} L \to L \otimes_{\mathcal{O}_X} P_{X/S}$ then maps $(a\otimes b) \otimes (c\otimes x)$ to $1\otimes x\otimes a\otimes bc$.

The above construction "is" Grothendieck's linearization of differential operators, if we apply it to $\mathcal{O}_S$-modules E which come from $\mathcal{O}_X$-module structures. Indeed, recall that in this case we can write $L(E) \underset{\text{def}}{=} \mathcal{O}_X \otimes_{\mathcal{O}_S} E \cong P_{X/S} \otimes_{\mathcal{O}_X} E$, through which <u>any</u> $\mathcal{O}_S$-linear map can be linearized – as we saw at the beginning of this chapter. The point of the above construction is that it furnishes $L(E)$ with a canonical stratification (which reduces to the usual one if $E = \mathcal{O}_S$).

If we take $E = \mathcal{O}_X$, then $L(E)$ is $P_{X/S}$, and the map $\theta: P_{X/S} \to P_{X/S} \otimes P_{X/S}$ maps $(c \otimes d)$ to $1 \otimes d \otimes 1 \otimes c$. This is a stratification of $P_{X/S}$ as an $\mathcal{O}_X$-module using the <u>left</u> structure, i.e., of $p_{1*}(P_{X/S})$; it is useful to notice that the map $\delta: P_{X/S} \to P_{X/S} \otimes P_{X/S}$ $c \otimes d \longmapsto c \otimes 1 \otimes 1 \otimes d$ is a stratification of $P_{X/S}$ using the <u>right</u> structure. (Indeed, if we had done the above construction using $E = K \otimes \mathcal{O}_X$ instead of $\mathcal{O}_X \otimes K$, we would have obtained δ .) Moreover, if $h: F \to G$ is a differential operator, and if we use the stratification δ to compute $\nabla(h): P \otimes F \to P \otimes G$, we find nothing other than $\bar{\delta}(h)$, immediately from the definitions. Note that $\nabla(h)$ is a differential operator if we use the right $\mathcal{O}_X$-module structure, but is $\mathcal{O}_X$-linear using the left $\mathcal{O}_X$-module structure. In fact it is better than that, it is compatible with the stratifications we have just put on $P \otimes F$ and $P \otimes G$. This comes from the commutativity of the diagram below:

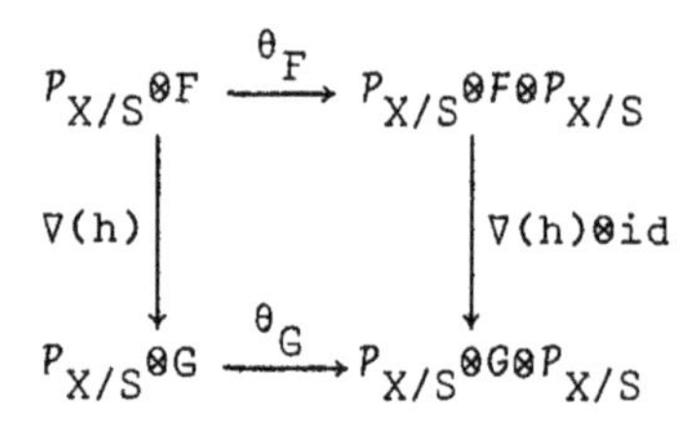

The above construction (really a complicated triviality) almost provides us with a way of going from the category of $\mathcal{O}_X$-modules and differential operators to the category of stratified $\mathcal{O}_X$-modules and horizontal $\mathcal{O}_X$-linear maps. I say "almost" because we have not passed to infinitesmal neighborhoods of the diagonal, which requires the replacement of $\mathcal{P}_{X/S}$ by the inverse system $\{\mathcal{P}^n_{X/S}\}$. The reader can check that θ_F induces maps $\mathcal{P}^{m+n}\otimes F \to \mathcal{P}^m \otimes F \otimes \mathcal{P}^n$ for all m and n, and that if $h: F \to G$ has order $\leq k$, $\nabla(k)$ induces a map $\mathcal{P}^m \otimes F \to \mathcal{P}^{m-k}\otimes G$ for all m. Passing to the limit, (or keeping the entire system in mind), we obtain Grothendieck's linearization of differential operators:

2.14 <u>Construction.</u> If F is an $\mathcal{O}_X$-module, let $L.(F)$ be the inverse limit (or system) of the $\mathcal{O}_X$-modules $\{\mathcal{P}^n \otimes F\}$. Then $L.(F)$ has a canonical stratification, and if $h: F \to G$ is a differential operator, we obtain a horizontal $\mathcal{O}_X$-linear, functorial map $L.(h): L.(F) \to L.(G)$. □

If X/S is smooth, we can use Construction (2.11) to obtain a sheaf on the infinitesmal site which is almost a crystal. The problem of the inverse limit is a nuisance, but it turns out that in positive characteristics we can avoid it. Therefore, we shall not pursue it further.

Let us now recall the obstructions to extending a connection $\nabla: E \to E \otimes \Omega^1_{X/S}$ to a stratification. Using the construction of 3bis) above, applied to the first order stratification ε_1 on E and the exterior derivative $d_k: \Omega^k_{X/S} \to \Omega^{k+1}_{X/S}$ (which is a differential operator of order ≤ 1), we get a map $\nabla^k: E \otimes \Omega^k \to E \otimes \Omega^{k+1}$. As expected, $\nabla^k(x \otimes \omega) = \nabla(x) \wedge \omega + x \otimes d_k \omega$.

It is not true, however, that $\nabla^{k+1}\circ\nabla^k = 0$ (since we have no cocycle condition), but it is well-known that the map $\nabla^1\circ\nabla = K: E \to E\otimes\Omega^2$ is $\mathcal{O}_X$-linear and that $(\nabla^{k+1}\circ\nabla^k)(x\otimes\omega) = K(x)\wedge\omega$. The map K is the "curvature" of ∇. In terms of local-coordinates $x_1 \dots x_n$, K has the following well-known expression: Let $\partial/\partial x_i \in \mathcal{D}er(\mathcal{O}_X,\mathcal{O}_X)$ map to $\theta_i \in \mathcal{D}iff^1(E,E)$ by ∇. Then the map $[\theta_i,\theta_j]$ $K: E \to E\otimes\Omega^2$ is given by $K = \sum_{i<j} [\theta_i,\theta_j]\otimes dx_i\wedge dx_j$. In particular, $K = 0$ iff the θ_i's <u>commute.</u> Thus we have the following well-known result:

2.15 <u>Theorem.</u> Suppose X/S is smooth and S is a $\mathbb{Q}$-scheme. Then a connection on E extends to a stratification iff $K=0$.

<u>Sketch.</u> *Certainly if* ∇ *extends to a stratification the* θ_i's commute, so that $K=0$. Conversely, given $\rho_1: \mathcal{D}iff^1(\mathcal{O}_X,\mathcal{O}_X) \to \mathcal{D}iff(E,E)$ we can extend it <u>uniquely</u> to $\rho: \mathcal{D}iff(\mathcal{O}_X,\mathcal{O}_X) \to \mathcal{D}iff(E,E)$, an $\mathcal{O}_X$-linear homomorphism, if $K=0$. To see this we can look locally, and hence can use the basis $\{D_q = \frac{1}{q!} D^{q_1} \dots D_n^{q_n}\}$ described in (2.6). Clearly the answer is forced on us: $\rho(D_q) = \frac{1}{q!}\theta_1^{q_1}\dots\theta_n^{q_n}$ – which exists because S is a $\mathbb{Q}$-algebra. This *makes* sense *because* the θ_i's commute, if $K=0$, so there is no ambiguity in the ordering.

Note that with no assumption on the characteristic, as soon as $K=0$ we get a complex $E\otimes\Omega^{\bullet}_{X/S}$, called the "de Rham complex of E". In characteristic zero one can apply Grothendieck's L construction to get a complex of pro-crystals on Inf(X/S), and Grothendieck proves that the cohomology of this complex is the Zariski hypercohomology of the complex $E\otimes\Omega^{\bullet}_{X/S}$. Our problem is to mimic this in characteristic p. First let us remind the reader that there are interesting connections in characteristic p with curvature 0 but which do <u>not</u> prolong to a stratification – namely the Gauss-Manin connection of a smooth map $X \to S$.

An elegant construction of the Gauss-Manin connection is explained, for instance, in [1] we shall not reproduce it. Grothendieck's argument that it does not, in general, extend to a stratification, is based on the following:

2.16 <u>Proposition.</u> Let k be a field, X a k-scheme of finite type, and $\phi: E \to F$ a horizontal morphism of coherent $\mathcal{O}_X$-module with stratification. Let $x \in X$ be a k-rational closed point. Then the stalk ϕ_x of ϕ at x vanishes iff the map $\phi(x): E\otimes k(x) \to F\otimes k(x)$ vanishes.

<u>Proof.</u> Let $\mathcal{O}$ be the local ring of X at x and m its maximal ideal. Of course, $\phi_x = 0$ iff $\phi_{x,n}: E/m^nE \to F/m^nF$ is zero for every n. Now the ring $\mathcal{O}_n = \mathcal{O}/m^n$ is Artinian, so the ideal defining the closed immersion $Spec\ \mathcal{O}_n \hookrightarrow Spec(\mathcal{O}_n)\times_k Spec(\mathcal{O}_n)$ is nilpotent. Thus, for N large enough, the map $\mathcal{O}_n\otimes\mathcal{O}_n \to P^N_{\mathcal{O}_n}$ is an isomorphism. Thus the formal descent data on E and F given us by the stratifications is <u>actual</u> descent data for E_n and F_n. Moreover, since $\mathcal{O}_n$ is faithfully flat over k, the descent data is <u>effective</u>, i.e. these exist k-modules E_0 and F_0 such that $E_n \cong E_0\otimes_k\mathcal{O}_n$ and $F_n \cong F_0\otimes_k\mathcal{O}_n$ for every n. Moreover, the

[1] Katz N. and Oda. "On the differentiation of DeRham cohomology classes with respect to parameters" <u>J. Math. Kyoto Univ.</u> <u>8</u> (1968). pp. 199-213.

horizontality of φ insures that there is also a k-linear map $\varphi_0 : E_0 \to F_0$ such that φ_n "is" $E_0 \otimes \mathcal{O}_n \xrightarrow{\varphi_0 \otimes id} F_0 \otimes \mathcal{O}_n$. Since φ_0 is necessarily $\varphi(x)$, $\varphi(x) = 0$ implies all $\varphi_n = 0$, as desired.

2.17 Note. The argument above also shows that a coherent sheaf with stratification on an algebraic X as above is necessarily locally free.

2.18 Example. Let $E \xrightarrow{f} S$ be a family of elliptic curves such that a finite number of the fibers have Hasse invariant zero. Let F_S: $S \to S$ be the absolute Frobenius morphism, and let $E^{(p^2)} \xrightarrow{g} S$ be the family obtained by pulling back E by F_S^2. The S-map $E \to E^{(p^2)}$ induced by F_E^2 induces a map of modules with integrable connection:

$$\mathbb{R}^1 g_*(\Omega^{\cdot}_{E^{(p^2)}/S}) \to \mathbb{R}^1 f_*(\Omega^{\cdot}_{E/S})$$

If we identify $\mathbb{R}^1 g_*(\Omega^{\cdot}_{E^{(p^2)}/S})$ with $\mathcal{O}_S \otimes_{F_S^2} \mathbb{R}^1 f_*(\Omega^{\cdot}_{E/S})$, this map becomes just the $\mathcal{O}_S$-linearization of the F_S^2-linear endomorphism of $\mathbb{R}^1 f_*(\Omega^{\cdot}_{E/S})$ induced by F_E^2 . Then it is easy to see that this map vanishes on a fiber E(s)/k(s) iff E(s) is supersingular, i.e. iff its Hasse invariant is zero. Thus, our proposition shows that $\mathbb{R}^1 f_*(\Omega^{\cdot}_{E/S})$ cannot be endowed with stratification in a functorial way. In particular, the Gauss-Manin connection does not prolong to a stratification. Actually,

N. Katz has given a far more profound explanation of this fact by relating the p-curvature of the Gauss-Manin connection to the Kodaira-Spencer map [2].

---

[2] Katz, N., "Algebraic Solutions of Differential Equations" Inv. Math. 18 (1972) 1-118.

§3. Divided Powers.

3.1 Definition. Let A be a commutative ring, $I \subset A$ an ideal. By "divided powers on I" we mean a collection of maps $\gamma_i : I \to A$, for all integers $i \geq 0$, such that:

1) For all $x \in I$, $\gamma_0(x) = 1$, $\gamma_1(x) = x$, $\gamma_i(x) \in I$ if $i \geq 1$.

2) For $x, y \in I$, $\gamma_k(x+y) = \sum_{i+j=k} \gamma_i(x)\gamma_j(y)$.

3) For $\lambda \in A$, $x \in I$, $\gamma_k(\lambda x) = \lambda^k \gamma_k(x)$.

4) For $x \in I$, $\gamma_i(x)\gamma_j(x) = ((i,j))\gamma_{i+j}(x)$, where

$$((i,j)) = \frac{(i+j)!}{(i!)(j!)} .$$

5) $\gamma_p(\gamma_q(x)) = C_{p,q}\gamma_{pq}(x)$, where $C_{p,q} = \dfrac{(pq)!}{p!(q!)^p}$.

Note. By induction on p it is simple to prove that $C_{p,q} = \prod_{i=1}^{p-1}((iq,q-1))$, and hence is an integer.[1]

We use the terminologies: "(I,γ) is a P.D. ideal", and "(A,I,γ) is a P.D. ring", and "γ is a P.D. structure on I".

Axioms 1) and 4) imply that $n!\gamma_n(x) = x^n$ for any $n \geq 0$, (by induction on n). Axiom 3) implies that $\gamma_k(0) = 0$ if $k > 0$.

A "P.D. morphism" $f:(A,I,\gamma) \to (B,J,\delta)$ is a ring homomorphism $f:A \to B$ such that $f(I) \subseteq J$ and such that $\delta_n(f(x)) = f(\gamma_n(x))$ for all n and all $x \in I$.

---

[1] In fact, $C_{p,q}$ is the number of partitions of a set with pq elements into p subsets with q each.

3.2 Examples.

1. $\{0\}$ is a P.D. ideal, with $\gamma_i(0) = 0$ for all $i \geq 1$.

2. If A is a $\mathbb{Q}$-algebra, every ideal has a unique P.D. structure, given of course by $\gamma_n(x) = x^n/n!$.

3. If V is a discrete valuation ring of unequal characteristic p and uniformizing parameter π , write $p = u\pi^e$, with u a unit. (e is called the absolute ramification index of V.) When does (π) have a P.D. structure? The answer is interesting: iff $e \leq p-1$. Clearly if γ exists it is unique, the problem is to determine when the elements $\gamma_n(x) = x^n/n!$ of the fraction field K of V lie in (π), for $x \in (\pi)$, and equally clearly it suffices to consider $x=\pi$. For this we must compute $\text{ord}_\pi(\pi^n/n!)$. We leave the proof of the following as an exercise:

3.3 Lemma. Let n be an integer, and write $n = \Sigma a_i p^i$, with $0 \leq a_i < p$. Then $\text{ord}_p(n!) = \frac{1}{p-1} \Sigma a_i(p^i-1)$. □

Thus, $\text{ord}_\pi(\gamma_n(\pi)) = n-\text{ord}_\pi(n!) = n-e\ \text{ord}_p(n!) =$

$$\Sigma a_i p^i - \frac{e}{p-1} \Sigma a_i(p^i-1) = \frac{1}{p-1} \Sigma a_i[p^i(p-1-e) + e] = \frac{p-1-e}{p-1} n+e \frac{\Sigma a_i}{(p-1)} .$$

Now $\gamma_n(\pi)$ lies in (π) for all $n \geq 1$ iff this expression is ≥ 1 for all choices of a_i, (not all zero), which clearly holds iff $p-1-e \geq 0$, i.e. iff $p-1 \geq e$. This proves our claim. Notice that in particular, if k is a field of characteristic $p > 0$, and V is a Cohen ring of k, $e = 1$, so (π) has a (unique) P.D. structure.

4. Suppose $mA = 0$. Then for all $x \in I$, $x^n = n!\gamma_n(x) = 0$, if $n \geq m$ so I is a nil ideal. If I can be generated by $\leq q$ elements, it follows that $I^{(m-1)q+1} = 0$, so I is a nilpotent ideal.

If $(m-1)!$ is invertible in A, and if $I^m = 0$, then I has a P.D. structure (<u>not</u> unique), e.g. given by $\gamma_m(x) = x^n/n!$ if $n < m$ and $= 0$ if $n \geq m$. In particular any ideal with $I^2 = 0$ has a P.D. structure, with $\gamma_n(x) = 0$ if $n \geq 2$.

In characteristic $p > 0$, any P.D. ideal satisfies $I^{(p)} = 0$ and any ideal with $I^p = 0$ has a P.D. structure. Here is an example, due to N. Koblitz, of an ideal with $I^{(p)} = 0$ but with nc P.D. structure (exercise): Take k a ring of characteristic p, $A = k[x_1 \ldots x_6]/(x_1^p \ldots x_6^p, x_1x_2 + x_3x_4 + x_5x_6)$ and $I = (x_1, \ldots x_6)$.

3.4 <u>Definition.</u> Let (I,γ) be a P.D. ideal. We say that an ideal $J \subseteq I$ is a "sub P.D. ideal" iff $\gamma_i(x) \in J$ for any $x \in J$, $i \geq 1$.

3.5 <u>Lemma.</u> If (A,I,γ) is a P.D. ring and $J \subseteq A$ is an ideal, then there is a P.D. structure $\bar{\gamma}$ (necessarily unique) on $I = I(A/J)$ such that $(A,I,\gamma) \to (A/J,\bar{I},\bar{\gamma})$ is a P.D. morphism, iff $J \cap I \subseteq I$ is a sub P.D. ideal.

<u>Proof.</u> If $\bar{\gamma}$ exists, it is clearly unique, and if $x \in J \cap I$, $n \geq 1$ $\bar{x} \in A/I$ is zero, so $0 = \bar{\gamma}_n(\bar{x}) = \overline{\gamma_n(x)}$, so $\gamma_n(x) \in I \cap J$. For the converse, we want to define $\bar{\gamma}_n(\bar{x}) = \overline{\gamma_n(x)}$ for any preimage x of $\bar{x}$. To see that it is

well-defined, we must check that if $y \in I \cap J$, $\gamma_n(x+y) \equiv \gamma_n(x)$ mod $I \cap J$. But $\gamma_n(x+y) = \sum_{i+j=n} \gamma_i(x)\gamma_j(y) = \gamma_n(x) + \sum_{j\geq 1} \gamma_{n-j}(x)\gamma_j(y)$ and the term $\sum_{j\geq 1}$ belongs to $I \cap J$ since it is a P.D. ideal. □

The following is useful when one is looking for sub-P.D. ideals. It, like the next lemma, is due to N. Roby, ["Les algèbres à puissances divisées" Bull. des Scie. Math. Fr. 2e sér., 89 (1965) p. 75-91].

3.6 Lemma. Let (A,I,γ) be a P.D. algebra, $S \subseteq I$ a subset, and $J \subseteq I$ the ideal generated by S. Then J is a sub-P.D. ideal of I iff $\gamma_n(s) \in J$ for each $s \in S$, and $n \geq 1$.

Proof. Necessity is obvious. For the converse, let J' be the subset of J consisting of just those x's for which $\gamma_n(x) \in J$ for all $n \geq 1$. Since $S \subseteq J' \subseteq J$, we have only to prove that J' is an ideal. If $x, y \in J'$ and $n \geq 1$, $\gamma_n(x+y) = \sum_{i+j=n} \gamma_i(x)\gamma_j(y)$, and since either i or j is ≥ 1 and J is an ideal, each term in the sum belongs to J, so $\gamma_n(x+y) \in J$ and $x+y \in J'$. If $x \in J'$ and $\lambda \in A$, $\gamma_n(\lambda x) = \lambda^n \gamma_n(x) \in J$, so $\lambda x \in J'$. □

Certain constructions can be carried out with P.D. structures without difficulty. For example, if $\{A_i, I_i, \gamma_i\}$ is a direct system of P.D. algebras and $A = \varinjlim A_i$, then $I = \varinjlim I_i$ has a unique P.D. structure γ such that each $(A_i, I_i, \gamma_i) \to (A, I, \gamma)$ is a P.D. morphism. For tensor products, a restriction is required:

3.7 <u>Lemma.</u> Suppose A is a ring, B and C are A-algebras, and $I \subseteq B$ and $J \subseteq C$ are augmentation ideals (i.e. there is a section of $B \to B/I$, etc.) with P.D. structures γ and δ, respectively. Then the ideal $K = \mathrm{Ker}\ (B \otimes_A C \to B/I \otimes C/J)$ has a unique P.D. structure ε such that $(B,I,\gamma) \to (B\otimes C,K,\varepsilon)$ and $(C,J,\delta) \to (B\otimes C,K,\varepsilon)$ are P.D. morphisms. □

We say that a sub-algebra B of (A,I,γ) is a "sub-P.D. algebra" iff for each $x \in I \cap B$, $\gamma_n(x) \in I \cap B$ for $n \geq 1$. Thus in this case there exists a (unique) P.D. structure on $I \cap B$ such that $(B,I \cap B,\delta) \to (A,I,\gamma)$ is a P.D. morphism. If I is an augmentation ideal there is a useful analogue of lemma (3.6), whose proof is so similar that we omit it:

3.8 <u>Lemma.</u> Let (A,I,γ) be a P.D. algebra, and assume $A \to A/I = A_0$ has a section s. Let $S_0 \subseteq A_0$ and $S_+ \subseteq I$ be subsets, and let B be the subring of A generated by $S = s(S_0) \cup S_+$. Then B is a sub P.D. algebra of (A,I,γ) iff $\gamma_n(s) \in B$ for every $s \in S_+$ and $n \geq 1$. □

Let us recall now the basic properties of the P.D. analogue of the symmetric algebra. Since its construction is fairly involved, we only sketch it in an appendix (following Roby [1,2]).

3.9 <u>Theorem.</u> Let M be an A-module. Then there exist a P.D. algebra $(\Gamma_A(M),\Gamma_A^+(M),\gamma)$ and an A-linear map $\varphi: M \to \Gamma_A^+(M)$ with the following universal property: If (B,J,δ) is any A-P.D.

algebra and $\psi: M \to J$ is a A-linear, there is a unique P.D. morphism $\bar{\psi}:(\Gamma_A(M), \Gamma_A^+(M),\gamma) \to (B,J,\delta)$ such that $\bar{\psi} \circ \varphi = \psi$. Moreover:

0) $\Gamma_A(M)$ is a graded algebra, $\Gamma_0(M) = A$, $\Gamma_1(M) = M$, and $\Gamma^+(M) = \bigoplus_{i \geq 1} \Gamma_i(M)$.

1) If A' is any A-algebra, $\Gamma_{A'}(M \otimes_A A') \cong A' \otimes_A \Gamma_A(M)$.

2) If $M = \varinjlim M_i$, $\Gamma_A(M) = \varinjlim \Gamma_A(M_i)$.

3) $\Gamma_A(M \oplus N) \cong \Gamma_A(M) \otimes \Gamma_A(N)$.

4) We use the notation $x^{[1]}$ for $\varphi(x)$ if $x \in M$, and $x^{[n]}$ for $\gamma_n(\varphi(x)) \in \Gamma_n(M)$. Sometimes we shall denote the divided power structure γ by [ ]. $\Gamma_n(M)$ is generated, as an A-module, by $\{x^{[q]} = x_1^{[q_1]} x_2^{[q_2]} \ldots x_k^{[q_k]} : \Sigma q_i = n, x_i \in M\}$. If $\{x_i : i \in I\}$ is a basis for M, $\{x^{[q]} : |q| = n, i \in I\}$. is a basis for $\Gamma_n(M)$.

If $\{x_i : i \in I\}$ is a basis for M, we shall also denote $\Gamma_A(M)$ by $A\langle x_i : i \in I\rangle$ and call it the P.D. polynomial A-algebra on the indeterminates $\{x_i : i \in I\}$. It has the expected universal mapping property with respect to P.D. algebras: If (B,J,δ) is any P.D. A-algebra and $y_i \in J$ for each $i \in I$, there is a unique P.D. homomorphism $A\langle x_i : i \in I\rangle \to (B,J,\delta)$ such that $x_i \to y_i$. □

It is now necessary to describe some other technical features of P.D. algebras. It would be too tedious to prove, or even state, all of them, so we only provide a sample.

3.10 Proposition. Let (I,γ) and (J,δ) be P.D. ideals in A. Then IJ is a sub-P.D. ideal of both I and J, and γ and δ agree on IJ.

Proof. IJ is generated (as an ideal) by the set of products xy: $x \in J$, $y \in J$. It is therefore enough to check that $\gamma_i(xy) = \delta_i(xy) \in IJ$ for $i,j \geq 1$. But $\gamma_i(xy) = y^i\gamma_i(x) \in IJ$, certainly. Moreover $\gamma_i(xy) = y^i\gamma_i(x) = i!\,\delta_i(y)\gamma_i(x)$ and by the same reasoning, $\delta_i(xy) = x^i\delta_i(y) = i!\,\gamma_i(x)\delta_i(y)$. □

3.11 Corollary. If I is a P.D. ideal, then $I^n \subseteq I$ is a sub-P.D. ideal, for all $n \geq 1$.

3.12 Proposition. Suppose (I,γ) and (J,δ) are P.D. ideals, suppose that $I \cap J$ is a sub-P.D. ideal of I and J, and suppose that γ and δ agree on $I \cap J$. Then there is a unique P.D. structure on $K = I+J$ such that I and J are sub-P.D. ideals.

Proof. We have an exact sequence of A-modules: $0 \to I \cap J \to I \oplus J \to I+J \to 0$. For each $x \in I$, let $g(x) = \sum_{n=0}^{\infty} \gamma_n(x)T^n$. Then $g: I \to \exp(A)$ is a homomorphism of A-modules, where $\exp(A) \subseteq A[[T]]$ is the A-module of power series of exponential type (c.f. appendix). Similarly, for $y \in J$, $d(y) = \sum_{n=0}^{\infty} \delta_n(y)T^n$ defines an A-linear $d: J \to \exp(A)$. We deduce that there is a unique A-linear $e: I+J \to \exp(A)$ which induces d and g. If $z \in I+J$, define $\varepsilon_n(z)$ by $\sum_0^{\infty} \varepsilon_n(z)T^n = e(z)$. It is easy to see that $\{\varepsilon_n\}$ satisfies the first four axioms for a divided power structure. For the last axiom we need a trick. Let M_X be the free A-module with basis X, M_Y the free A-module with basis Y, and $M = M_X \oplus M_Y$. For any $z \in I+J$, write $z = x+y$, with $x \in I$, $y \in J$, and consider the P.D. morphisms $\Gamma(M_X) \to (A,I,\gamma)$ and $\Gamma(M_Y) \to (A,J,\delta)$ sending $X \mapsto x$, $Y \mapsto y$, respectively. We deduce an algebra homomorphism $\Gamma(M_X)\otimes\Gamma(M_Y) \to A$. But $\Gamma(M_X)\otimes\Gamma(M_Y) \cong \Gamma(M)$ and hence is a P.D. algebra, and the element $Z = X+Y$ maps to z. Since in $\Gamma(M)$ axiom 5 holds, i.e. $(Z^{[n]})^{[i]} = C_{i,n}Z^{[in]}$, it also holds in A. □

3.13 <u>Definition.</u> Suppose (A,I,γ) is a P.D. ring and $S \subseteq I$ is a subset. We say that S is a "set of P.D. generators of I" iff I is the smallest sub P.D. ideal of I containing S — equivalently, iff I is generated as an ideal by $\{\gamma_j(s):j\geq 1\}$.

We shall often be working with algebras over some fixed P.D. ring as "base", usually a truncated Witt ring $(W_m,(p),\gamma)$. The ideal $(p) \subseteq W_m$ has many P.D. structures; we shall work with the "canonical" one, induced from the <u>unique</u> P.D. structure on $(p) \subseteq W_\infty$, using (3.5) and (3.11). We want to consider thickenings (P,J,δ) of W/pW algebras, where the P.D. strucutre δ on J is <u>compatible</u> with the canonical one on $(p) \subseteq W_m$. Since $p \in J$ this just says that $(W_m,(p),\gamma) \to (B,J,\delta)$ is a P.D. morphism. It is convenient, however, to have a more general notion of compatibility which does not require that the P.D. ideals be preserved.

3.14 <u>Definition.</u> Let (A,J,γ) be a P.D. ring and B an A-algebra. We say that " γ extends to B" iff there is a P.D. structure $\bar{\gamma}$ on IB such that $(A,I,\gamma) \to (B,IB,\bar{\gamma})$ is a P.D. morphism.

<u>Notes.</u>

1) If $\bar{\gamma}$ exists it is unique. This is easy.

2) γ extends to B iff thereis an P.D. ideal (J,δ) of B such that $(A,I,\gamma) \to (B,J,\delta)$ is a P.D. morphism, for if such a map exists, it is easy to see using Lemma (3.6) that IB is a sub P.D. ideal of J.

3) In general $\bar{\gamma}$ does not exist; for instance if $B = A/J$ and $I \cap J \subsetneq I$. is not a sub P.D. ideal.

3.15 <u>Proposition.</u> Suppose I is principal. Then γ extends to any B.

<u>Proof.</u> In this case $IB = \{tb \in B\}$ for some fixed $t \in I$, and we want to define $\bar{\gamma}_n(tb) = b^n\gamma_n(t)$. In fact this is well-defined, because if $tb = tb'$, $b^n\gamma_n(t) - b'^n\gamma_n(t) =$
$\sum_{i=1}^{n-1} b^i b'^{n-i-1}(b-b')\gamma_n(t)$. But $\gamma_n(t)$ is a multiple of t, so $(b-b')\gamma_n(t) = 0$ also. This shows that our definition makes sense, and it is easy to see that it is a P.D. structure. □

3.16 <u>Proposition.</u> Let (A,I,γ) be a P.D. ring, B an A-algebra, (J,δ) a P.D. ideal in B. Then the following are equivalent:

(1) γ extends to B and $\bar{\gamma} = \delta$ on $IB \cap J$.

2) The ideal $K = IB+J$ has a (necessarily unique) P D structure $\bar{\delta}$ such that $(A,I,\gamma) \to (B,K,\bar{\delta})$ and $(B,J,\delta) \to (B,K,\bar{\delta})$ are P.D. morphisms.

3) There is an ideal $K \supseteq IB+J$ with a P.D structure δ' such that $(A,I,\gamma) \to (B,K',\delta')$ and $(B,J,\delta) \to (B,K',\delta')$ are P.D. morphisms. □

Proof. (1) → (2) is just (3.12). (2) → (3) is trivial. For (3) → (1) observe that γ extends to B by note 2 above and that $\bar{\gamma} = \delta$ on $IB \cap J$ because the two maps are P. D. morphisms. □

3.17 Definition. If the equivalent conditions of the above proposition are fulfilled, we say that γ and δ are "compatible".

3.18 Remark. If B is an augmented A-algebra with a P. D. augmentation ideal (J,δ) and $B/J \cong A$, then δ is compatible with any P. D, structure on any ideal I of A. To see this, let $K = IB+J$ and observe that since $B = A\oplus J$, $K = I\oplus J$, which has a P.D. structure, as is easy to see.

We are now ready to construct one of the divided power analogues of formal completion, namely, the "P.D. envelope" of an ideal. We work systematically over a fixed P. D. algebra (A,J,γ) and consider only PD structures compatible with γ.

3.19 Theorem. Let (A,I,γ) be a P. D. algebra and let J be an ideal in an A-algebra B. Then there exists a B-algebra $\mathcal{D}_{B,\gamma}(J)$

with a P.D. ideal $(\bar{J},[\ ])$, such that $J\mathcal{D}_{B,\gamma}(J) \subseteq \bar{J}$, such that $[\ ]$ is compatible with γ, and with the following universal property: For any B-algebra C containing an ideal K which contains JC and with a P.D. structure δ compatible with γ, there is a unique P.D. morphism $(\mathcal{D}_{B,\gamma}(J), \bar{J},[\ ]) \longrightarrow (C,K,\delta)$ making the diagram commute:

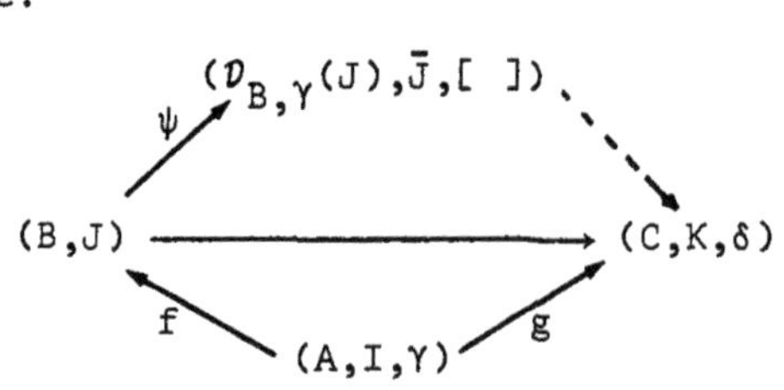

Proof. First we do the construction in the special case:

Case 1. $f(I) \subseteq J$. (In this case, g and $\psi \circ f$ will also be P.D. morphisms.)

The construction in this case is as follows: Start with the P.D. algebra $(\Gamma_B(J),\Gamma_B^+(J),[\ ])$ of Theorem (3.9), with $\varphi: J \to \Gamma_1(J)$ the universal map. Consider the ideal $\mathcal{J}$ of $\Gamma_B(J)$ generated by elements of the following two forms:

(i) $\varphi(x)-x$ for $x \in J$.

(ii) $\varphi(f(y))^{[n]} - f(\gamma_n(y))$ for $y \in I$.

Claim. $\mathcal{J} \cap \Gamma_B^+(J)$ is a sub PD ideal of $\Gamma_B^+(J)$. To prove this, note first that we get the same $\mathcal{J}$ if we replace relation (ii) by:

(ii)' $\varphi(f(y))^{[n]} - \varphi(f(\gamma_n(y)))$ for $y \in I$.

Let $\mathcal{J}_1$ be the ideal generated by the elements of the form (i) and $\mathcal{J}_2$ the ideal generated by elements of the form (ii)', so that $J = J_1 + J_2$. Since $J_2 \subseteq \Gamma_B^+(J)$, $J \cap \Gamma_B^+(J) = J_1 \cap \Gamma_B^+(J) + J_2$. Thus, using the formula for $\gamma_n(x+y)$, we see that it suffices to show that $x^{[n]} \in J \cap \Gamma_B^+(J)$ if x belongs either to $J_1 \cap \Gamma_B^+(J)$ or to J_2.

First suppose $x \in J_1 \cap \Gamma_B^+(J)$, say $x = \Sigma a_i(\varphi(x_i) - x_i))$, with $a_i \in \Gamma_B(J)$. Since $x \in \Gamma_B^+(J)$, the degree zero part of this sum is zero. Thus, if we write $a_i = a_i^0 + a_i^+$, with $a_i^0 \in B$ and $a_i^+ \in \Gamma_B^+(J)$, we have $\Sigma a_i^0 x_i = 0$. It follows that $0 = \varphi(\Sigma a_i^0 x_i) = \Sigma a_i^0 \varphi(x_i)$, and hence that $x = \Sigma a_i^+(\varphi(x_i) - x_i) \in I_1 \Gamma_B^+(J)$, in fact. In other words, $J_1 \cap \Gamma_B^+(J) = J_1 \Gamma_B^+(J)$, which is easily seen to be a sub PD ideal of $\Gamma_B^+(J)$, by (3.6).

Now suppose $x \in J_2$. By (3.6) it suffices to see that if $y \in I$ and $m \geq 1$, $[\varphi(f(y))^{[n]} - \varphi(f(\gamma_n(y)))]^{[m]} \in J \cap \Gamma^+(J)$. Obviously it belongs to $\Gamma^+(J)$; in the remaining computations, we shall write $x^{[n]}$ for $\varphi(x)^{[n]}$. Compute:

$$(f(y)^{[n]} - f(\gamma_n(y))^{[1]})^{[m]} = \sum_{r+s=m} (f(y)^{[n]})^{[r]} (f(\gamma_n(y))^{[1]})^{[s]} (-1)^s$$

$$= \sum_{r+s=m} (-1)^s C_{r,n} f(y)^{[nr]} f(\gamma_n(y))^{[s]}.$$ From the definition of J, we see that this is congruent mod J to

$$\sum_{r+s=m} (-1)^s C_{r,n} f(\gamma_{nr}(y)) f(\gamma_s \gamma_n(y)) = f\Big(\sum_{r+s=m} (-1)^s C_{r,n} \gamma_{nr}(y) \gamma_s(\gamma_n(y))\Big)$$

$$= f\Big(\sum_{r+s=m} (-1)^s \gamma_r(\gamma_n(y)) \gamma_s(\gamma_n(y))\Big) = f((\gamma_n(y) - \gamma_n(y))^{[m]}) = 0.$$

This completes the proof that $J \cap \Gamma^+$ is a sub P.D. ideal and allows us to conclude that the image $\bar{J}$ of $\Gamma_B^+(J)$ in $\mathcal{D} = \Gamma_B(J)/I$ has a P.D. structure, which we also denote by [ ]. The first set of relations in J imply that $J\mathcal{D} \subseteq \bar{J}$, and the second set insures that [ ] is compatible with γ. For the time being we verify the universality with the additional restriction that $IC \subseteq K$. If this is the case, since $JC \subseteq K$, we get a P.D. morphism $(\Gamma_B(J), \Gamma^+, [\ ]) \xrightarrow{\mu} (C,K,\delta)$ inducing the map $J \to K$, and since $(A,J,\gamma) \to (C,K,\delta)$ is also a P.D. morphism, it is easy to see that this map μ factors through $\mathcal{D}$.

The General Case. Let $J_1 = J+IB$, $\mathcal{D}_{B,\gamma}(J) = \mathcal{D}_{B,\gamma}(J_1)$, and let $\bar{J} \subseteq \bar{J}_1$ be the sub P.D. ideal generated by J (i.e. the ideal generated by $\{x^{[n]} : n \geq 1, x \in J\mathcal{D}\}$. If (C,K,δ) is as in the theorem, let $K_1 = K+IC$ with its P.D. structure δ'; then by case 1 there is a P.D. map $(\mathcal{D},\bar{J}_1,\bar{\gamma}) \to (C,K_1,\delta')$. Since $\bar{J}C$ is the sub P.D. ideal generated by JC, it is contained in K. □

3.20 Remarks.

1) By the very construction of $\mathcal{D}$, we see that $\mathcal{D}_{B,\gamma}(J) = \mathcal{D}_{B,\gamma}(J+IB)$, and hence that the algebra $\mathcal{D}$ depends only on J+IB. Of course, the P.D. ideal $\bar{J}$ still depends on J.

2) If the structure map $A \to B$ factors through some A', and if γ extends to γ' on A', then $\mathcal{D}_{B,\gamma}(J) = \mathcal{D}_{B,\gamma'}(J)$. This is because in the set of generators of type ii, it suffices to consider y's from any generating set of I, or of IA'.

3) As a B algebra, $\mathcal{D}_{B,\gamma}(J)$ is generated by $\{x^{[n]}: n \geq 0, x \in J\}$, since this was true already for $\Gamma_B(J)$. Moreover, any set of generators of J gives us a set of P.D. generators for $\bar{J}$.

4) In general it is not true that $B/J \cong \mathcal{D}_{B,\gamma}(J)/\bar{J}$, since such an equality would imply that γ extends to B/J. Conversely, if γ extends, we get equality, because the universal mapping property tells us that there is a map $\mathcal{D}_{B,\gamma}(J) \to B/J$ sending $\bar{J}$ to zero, which then induces an inverse to the canonical $B/J \to \mathcal{D}_{B,\gamma}(J)/\bar{J}$. Note that γ automatically extends if I is principal (by (3.15)), or if $IB \subseteq J$ (trivial).

5) If M is an A-module, if $B = \mathrm{Sym}^{\cdot}_A(M)$, and if J is the ideal $\mathrm{Sym}^{+}_A(M)$, then $\mathcal{D}_{S^{\cdot}(M),0}(J) = \Gamma_A(M)$. (When we write $\gamma = 0$, we mean with the trivial P.D. structure on the zero ideal of A.) This is easy to check from the universal mapping properties. In particular, if $B = A[X_1 \dots X_n]$ and $J = (X_1 \dots X_n)$, then $\mathcal{D}_{B,0}(J)$ is the P.D. polynomial algebra $A\langle X_1 \dots X_n\rangle$.

6) Suppose that γ extends to B/J and in addition that $B \to B/J$ has a section. Then we can drop the compatibility conditions, i.e. $\mathcal{D}_{B,\gamma}(J) \cong \mathcal{D}_{B,0}(J)$. To see this, let $\bar{\gamma}$ denote the extension of γ to B/J. The section $B/J \to B \to D_{B,0}(J)$ of $D_{B,0}(J) \to D_{B,0}(J)/\bar{J} \cong B/J$ allows us to write $D_{B,0}(J) = B/J \oplus \bar{J}$. Then $\bar{\gamma}$ extends to $D_{B,0}(J)$, and so by the universal mapping property, we get a map $D_{B,\gamma}(J) \to D_{B,0}(J)$, inverse to the canonical surjective map $D_{B,0}(J) \to D_{B,\gamma}(J)$.

7) If K is an ideal of B such that $KD_{B,\gamma}(J) = 0$, then $\mathcal{D}_{B,\gamma}(J) \cong \mathcal{D}_{B/K,\gamma}(J/J \cap K)$. This is an exercise in universal mapping properties. An example arises when m is an integer such that $mB = 0$ and J has $\leq q$ generators.

Then $J^{(m-1)q+1}\mathcal{D}_{B,\gamma}(J) = 0$ and hence $\mathcal{D}_{B,\gamma}(J) \cong \mathcal{D}_{B/J^{(m-1)q+1}}(J/J^{(m-1)q+1})$. Thus $\mathcal{D}_{B,\gamma}(J)$ depends only on an infinitesimal neighborhood of $V(J)$ in *Spec* B.

8) Suppose that $(A,I,\gamma) \to (A',I',\gamma')$ is a surjective P.D. morphism, and $B' = A'\otimes_A B$ $J' = JB'$. Then the canonical map: $A'\otimes_A \mathcal{D}_{B,\gamma}(J) \twoheadrightarrow \mathcal{D}_{B',\gamma'}(J')$ is an isomorphism. It suffices to see that the image J'' of $\bar{J}$ in $A'\otimes_A \mathcal{D}_{B,\gamma}(J)$ has a P.D. structure compatible with γ (c.f. Remark 1), and hence it suffices to see that the kernel of $\mathcal{D}_{B,\gamma}(J) \to A'\otimes_A \mathcal{D}_{B,\gamma}(J)$ meets $\bar{J}$ in a sub P.D. ideal. But if $A' = A/K$, this kernel is just $K\mathcal{D}_{B,\gamma}(J)$, and since $K \subseteq I$ is a sub P.D. ideal and $[\ ]$ is compatible with γ, this is clear from (3.16).

3.21 Proposition. Suppose J is an ideal in the (A,I,γ) algebra B and B' is a B algebra. Then there is a natural map $\mathcal{D}_{B,\gamma}(J)\otimes_B B' \to \mathcal{D}_{B',\gamma}(JB')$, which is an isomorphism if B' is flat over B.

Proof. The map comes from the map $\mathcal{D}_{B,\gamma}(J) \to \mathcal{D}_{B',\gamma}(JB')$. In the flat case, $JB' \cong J\otimes_B B'$ so $\Gamma_{B'}(JB') \cong \Gamma_B(J)\otimes_B B'$ is flat over $\Gamma_B(J)$. From the description of $\mathcal{J}$, we see easily that $\mathcal{J}' = \mathcal{J}\Gamma_{B'}(JB') \cong \mathcal{J}\otimes\Gamma_B(J)\otimes B' \cong \mathcal{J}\otimes B'$, so $\mathcal{D}_{B',\gamma}(JB') = \Gamma_{B'}(JB')/\mathcal{J}'$ is isomorphic to $\Gamma_B(J)\otimes B'/\mathcal{J}\otimes B' \cong \mathcal{D}_{B,\gamma}(J)\otimes_B B'$. □

3.22 <u>Corollary.</u> If B is a flat (A,I,γ)-algebra, γ extends to B.

<u>Proof.</u> In general, to give a P.D. structure on IB compatible with γ is equivalent to giving a section of the canonical $B \to \mathcal{D}_{B,\gamma}(IB)$ such that if K is its kernel, $K \cap \overline{IB}$ is a sub P.D. ideal of $\overline{IB}$. In particular we have a map $\mathcal{D}_{A,\gamma}(I) \to A$ with a P.D. kernel, and hence a map $\mathcal{D}_{A,\gamma}(I) \otimes B \to B$. By flatness, this is a map $\mathcal{D}_{B,\gamma}(IB) \to B$, and it is easy to check that $K \cap \overline{IB} \subseteq \overline{IB}$ is a sub P.D. ideal. □

3.23 <u>Corollary.</u> The map $B \to \mathcal{D}_{B,\gamma}(J)$ is an isomorphism mod $\mathbb{Z}$-torsion.

<u>Proof.</u> Let $B' = B \otimes_{\mathbb{Z}} \mathbb{Q}$. The map $B \to B'$ is flat, and of course $\mathcal{D}_{B',\gamma}(JB') \cong B'$. But $\mathcal{D}_{B',\gamma}(JB') \cong \mathcal{D}_{B,\gamma}(J) \otimes_B B' \cong \mathcal{D}_{B,\gamma}(J) \otimes_{\mathbb{Z}} \mathbb{Q}$. Thus the map $B \to \mathcal{D}_{B,\gamma}(J)$ becomes an isomorphism when tensored with $\mathbb{Q}$. □

3.24 <u>Definition.</u> Let (A,I,γ) be a P.D. ring, $n \geq 1$ an integer. Then $I^{[n]}$ is the ideal generated by $\{\gamma_{i_1}(x_1)\gamma_{i_2}(x_2)\ldots\gamma_{i_k}(x_k) :$ $\Sigma i_j \geq n$ and $x_j \in I\}$.

3.25 <u>Proposition.</u> $I^{[n]} \subseteq I$ is a sub P.D. ideal, and $I^{[n]}I^{[m]} \subseteq I^{[n+m]}$.

<u>Proof.</u> Compute $\gamma_p(\gamma_{i_1}(x_1)\ldots\gamma_{i_k}(x_k)) =$ $\gamma_{i_1}(x_1)^p\ldots\gamma_{i_{k-1}}(x_{k-1})^p\gamma_p(\gamma_{i_k}(x_k)) = N\gamma_{pi_1}(x_1)\ldots\gamma_{pi_k}(x_k)$ for

some integer N, and hence belongs to $I^{[n]}$, if $p \geq 1$. The next statement is obvious. □

3.26 <u>Warning.</u> $I^{[n]}$ is <u>not</u> generated by $\{\gamma_{i_1}(x_1)\ldots\gamma_{i_k}(x_k) : \Sigma i_j = n\}$, in general. For example if (A,I,γ) is the Witt ring with its canonical structure, $p^{[n]} = p^n/n!$, and the ideal $(p)^{[n]}$ is (p^ν) where $\nu = \inf\{\nu_p(p^k/k!) : k \geq n\}$. Since the sequence $\nu_p(p^k/k!)$ is not monotone increasing, this is not just $\nu_p(p^n/n!)$, in general.

3.27 <u>Definition.</u> A P.D. ideal I is "P.D. nilpotent" iff $I^{[n]} = 0$ for some $n \geq 1$.

In general, if I is P.D. nilpotent it is nilpotent, but not conversely; for example, take the ideal (2) in $\mathbb{Z}/2^m\mathbb{Z}$ with $m > 1$.

3.28 <u>Proposition.</u> Let V be a discrete valuation ring with parameter π, $p = a\pi^e$. Recall that (π) has a (unique) divided power structure iff $e \leq p-1$. This structure induces a <u>nilpotent</u> P.D. structure on $V/(\pi^m V)$ iff $e < p-1$ (where $m > 1$).

This proposition follows easily from the formula for $(\pi^k/k!)$ given in (3.3).

The notion of nilpotent P.D. structure gives rise to another notion of P.D. envelope which is useful for some purposes.

3.29 <u>Definition.</u> If $J \subseteq B$ is an ideal and $n \geq 0$ is an integer, $\mathcal{D}^n_{B,\gamma}(J) = \mathcal{D}_{B,\gamma}(J)/\bar{J}^{[n+1]}$, and $\mathcal{D}^\infty_{B,\gamma}(J) = \varprojlim \mathcal{D}^n_{B,\gamma}(J)$.

We can sheafify the notion of a P.D. algebra, and speak of a sheaf of P.D. rings $(\mathcal{A},\mathcal{I},\gamma)$ on a space X (for now a topological space, later a topos), meaning the obvious things. If $f:X \to Y$ is a map, $(f_*\mathcal{A},f_*\mathcal{I},f_*\gamma)$ is a sheaf of P.D. rings on Y, and if $(\mathcal{B},\mathcal{J},\delta)$ is a sheaf of P.D. rings on Y, $(f^{-1}(\mathcal{B}), f^{-1}(\mathcal{J}), f^{-1}(\gamma))$ is a sheaf of P.D. rings on X. A "P.D. ringed space" is a pair $(X,(\mathcal{A},\mathcal{I},\gamma))$ where X is a space and $(\mathcal{A},\mathcal{I},\gamma)$ is a sheaf of P.D. rings on X. A morphism of P.D. ringed spaces is a continuous map $f:X \to Y$ together with a map sheaves of P.D. rings: $(\mathcal{B},\mathcal{J},\delta) \to (f_*\mathcal{A},f_*\mathcal{I},f_*\gamma))$ (in particular, $\mathcal{J} \to f_*\mathcal{I}$).

If (A,I,γ) is a P.D. algebra and $\alpha \in A$, it is easy to see that the localization (A_α,I_α) has a canonical P.D. structure γ such that $(A,I,\gamma) \to (A_\alpha,\gamma\ )$ is a P.D. morphism: just set $\gamma_n(x/\alpha^i) = \gamma_n(x)/\alpha^{in}$. (In fact we have already seen in (3.22) that γ extends to any flat A-algebra). Thus we get a sheaf of P.D. algebras on the spectrum of A, and hence we can regard $\mathcal{S}pec(A,I,\gamma)$ as a P.D. ringed space. Moreover we can reverse the procedure: If $X = \mathcal{S}pec\ A$ and $\mathcal{I} \subseteq \mathcal{O}_X$ is a quasi-coherent sheaf of ideals, one sees easily by taking global sections that P.D. structures on $\mathcal{I}$ correspond to P.D. structures on $I = \Gamma(X,\mathcal{I})$. Similarly, the P.D. morphisms $\mathcal{S}pec(A,I,\gamma) \to \mathcal{S}pec(B,J,\delta)$ can be identified with P.D. morphisms $(B,J,\delta) \to (A,J,\gamma)$. The following result follows easily from (3.21):

3.30 <u>Proposition.</u> Let S be a scheme, $\mathcal{I} \subseteq \mathcal{O}_S$ a quasi-coherent sheaf of ideals with a P.D. structure γ, and let X be an S-scheme.

Then if $\mathcal{B}$ is a quasi-coherent $\mathcal{O}_X$-algebra and $\mathcal{J} \subsetneq \mathcal{B}$ is a quasi-coherent ideal, $\mathcal{D}_{\mathcal{B},\gamma}(\mathcal{J})$ is a quasi-coherent $\mathcal{O}_X$-algebra. □.

In the discussion which follows let us fix (S,I,γ) as in the proposition, and suppose $i:X \to Y$ is a closed immersion of S-schemes. We use the notation $D_{X,\gamma}(Y)$ for $\mathcal{D}_{\mathcal{O}_{Y,\gamma}}(\mathcal{J})$, if $\mathcal{J}$ defines $X \subseteq Y$, and because of the proposition, we can define a scheme $D_{X,\gamma}(Y) = \mathit{Spec}_Y(\mathcal{D}_{X,\gamma}(Y))$.

If γ extends to X (i.e. to $\mathcal{O}_X$), then $\mathcal{D}_{X,\gamma}(Y)/\bar{\mathcal{J}} \cong \mathcal{O}_X$, as we have seen, so that $i:X \to Y$ factors through a closed *immersion* $j:X \to D_{X,\gamma}(Y)$ *with kernel* $\bar{\mathcal{J}}$ *a P.D. ideal (we say* that j is a P.D. immersion). The P.D. immersion j is universal: if $i':X' \to Y'$ is a P.D. immersion (compatible with γ) and if the solid diagram below exists, we get a unique $Y' \to D_{X,\gamma}(Y)$ as shown:

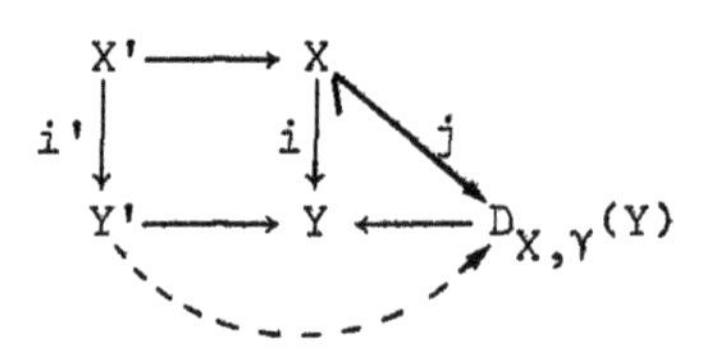

3.31 <u>Remark.</u> If $X \subseteq Y$ is only locally closed (and γ extends to X) then $\mathcal{D}^n_{X,\gamma}(Y) = \mathcal{D}_{X,\gamma}(Y)/\bar{\mathcal{J}}^{[n+1]}$ still makes sense, as does $\mathcal{D}_{X,\gamma}(Y)$ if $m\mathcal{O}_Y = 0$ for some m. This is because $\bar{\mathcal{J}}\mathcal{D}^n$(resp. $\bar{\mathcal{J}}\mathcal{D}$) is a nil ideal, so that the underlying topological space of D^n(resp. D) is the same as that of X, and we can therefore replace Y by an open neighborhood of X in which it is closed.

We call $D^n_{X,Y}(Y)$ the nth order divided power-neighborhood of X in Y (even though it is not a subscheme of Y), and $D_{X,Y}(Y)$ the divided power envelope or neighborhood of X in Y. We can compute it, locally, in the following case:

3.32 <u>Proposition.</u> Suppose $X \to Y$ is an immersion of smooth S-schemes and $m\mathcal{O}_Y = 0$. Then $\mathcal{D}_{X,Y}(Y)$ is locally isomorphic to a P.D. polynomial algebra over $\mathcal{O}_X$.

<u>Proof.</u> By the above remark, we may assume that $X \subseteq Y$ is closed, say of codimension d, and defined by the ideal $\mathcal{J}$. Since is locally generated by d sections and $m\mathcal{O}_Y = 0$, $\mathcal{J}^N\mathcal{D} = 0$, where $N = (m-1)d+1$, and hence by (3.20.7), $\mathcal{D}_{\mathcal{O}_{Y},Y}(\mathcal{J}) = \mathcal{D}_{\mathcal{O}_Y/\mathcal{J}^N,Y}(\mathcal{J}/\mathcal{J}^N)$. Since X/S is smooth, the map $\mathcal{O}_Y/\mathcal{J}^N \to \mathcal{O}_X$ locally has a section, and γ extends to X, so we may drop the subscript γ by (3.20.6). Let $t_1 \ldots t_d$ be a regular sequence of sections of $\mathcal{J}$ which generate $\mathcal{J}$; using the t's and the section we get a map $\mathcal{O}_X[t_1 \ldots t_d] \to \mathcal{O}_Y$ (locally, of course). If $\mathcal{J}_0 \subseteq \mathcal{O}_X[t_1 \ldots t_d]$ is the ideal $(t_1 \ldots t_d)$, this map induces an isomorphism: $\mathcal{O}_X[t_1 \ldots t_d]/\mathcal{J}_0^N \cong \mathcal{O}_Y/\mathcal{J}^N$. Thus we have:

$$\mathcal{D}_{X,Y}(Y) = \mathcal{D}_{\mathcal{O}_{Y},Y}(\mathcal{J}) \cong \mathcal{D}_{\mathcal{O}_Y/\mathcal{J}^N,Y}(\mathcal{J}/\mathcal{J}^N)$$

$$\cong \mathcal{D}_{\mathcal{O}_Y/\mathcal{J}^N,0}(\mathcal{J}/\mathcal{J}^N) \cong \mathcal{D}_{\mathcal{O}_X[t_1 \ldots t_d]/\mathcal{J}_0^N,0}(\mathcal{J}_0/\mathcal{J}_0^N)$$

$$\cong \mathcal{D}_{\mathcal{O}_X[t_1 \ldots t_d],0}(\mathcal{J}_0) \cong \mathcal{O}_X\langle t_1 \ldots t_d\rangle \ . \ \square$$

For many purposes it is convenient to work over a formal base, e.g., a p-adic base. Hence we shall need to discuss some compatibilities of the constructions of this chapter with inverse limits.

Let (A,I,γ) be a Noetherian P.D. ring with $P \subseteq I$ a sub P.D.-ideal. (The most important case is A = the Witt ring of a perfect field and $P = I = (p)$, with γ its unique P.D. structure.) Recall that $P^{n+1} \subseteq I$ is a sub P.D. ideal, so that we have a natural P.D. morphism $(A,I,\gamma) \to (A_n,I_n,\gamma) \underset{\text{def}}{=} (A/P^{n+1},IA/P^{n+1},\gamma)$. It is easy to see that the operations γ on I_n induce a P.D. structure on $\hat{I} = \varprojlim I_n \subseteq \varprojlim A_n$, so that we have P.D. morphisms:

$$(A,I,\gamma) \to (\hat{A},\hat{I},\gamma) \to (A_n,I_n,\gamma) \qquad \textit{for all } n .$$

Now assume that A is P-adically complete. For any formal A-scheme Z, let Z_n denote $Z \times_{\mathit{Spec}\, A} \mathit{Spec}\, A_n$.

3.33 <u>Proposition.</u> Suppose Y is a formal A-scheme with ideal of definition containing $P\mathcal{O}_Y$, and assume that γ extends to Y. If $J \subseteq \mathcal{O}_Y$ is a sheaf of ideals, there are canonical isomorphisms:

$$A_n \otimes_A \mathcal{D}_{\mathcal{O}_Y,\gamma}(J) \to \mathcal{D}_{\mathcal{O}_{Y_n},\gamma}(J\mathcal{O}_{Y_n}) \qquad \text{and}$$

$$\hat{\mathcal{D}}_{\mathcal{O}_Y,\gamma}(J) \to \varprojlim_n \mathcal{D}_{\mathcal{O}_{Y_n},\gamma}(J\mathcal{O}_{Y_n}) ,$$

where the ^ means P-adic completion. Moreover, $\hat{\bar{J}}$ has a canonical P.D. structure compatible with γ .

Proof. The first statement is an immediate consequence of (3.20.8), and the rest follows immediately. □

3.34 Corollary. Let $X \subseteq Y$ be a formal subscheme, and let $Y_{/X}$ denote the formal completion of Y along X. Assume that the ideal P contains a nonzero integer. Then there is a canonical isomorphism:

$$\hat{D}_{X,\gamma}(Y_{/X}) \to \hat{D}_{X,\gamma}(Y) .$$

Proof. According to the previous result, it suffices to prove this over A_n instead of A. But there we can appeal to (3.20.7), exactly as we did in the proof of (3.32). □

3.35 Corollary. Suppose that Y/A and X/A_0 smooth. Then $\hat{D}_{X,\gamma}(Y)$ is locally isomorphic to the P-adic completion of a P.D. polynomial algebra with coefficients in a formally smooth A-algebra. In particular, if A has no $\mathbb{Z}$-torsion, $\mathcal{O}_{\hat{D}_{X,\gamma}(Y)}$ has none.

Proof. Locally it is easy to find a closed $Z \hookrightarrow Y$ formally smooth over A such that $Z_0 = X$. Recall from (3.20.1) that $D_{X,\gamma}(Y) = D_{Z,\gamma}(Y)$. Now the formal completion of Y along Z is locally isomorphic to the formal completion of $\mathbb{A}^n_Z$ along Z, so by (3.34) it suffices to consider this case – which is trivial.

If A has no $\mathbb{Z}$-torsion, we can be quite explicit. If C is a formally smooth (hence flat) A-algebra, $C\langle t_1 \dots t_n\rangle$ is easily seen to be the C-subalgebra of $\mathbb{Q}\otimes_{\mathbb{Z}} C[t_1 \dots t_n]$ generated by all the elements $t_i^{[k]} = t_i^k/k!$. Clearly any element of $C\langle t_1 \dots t_d\rangle$

can be uniquely written as a polynomial $\Sigma a_k t^{[k]}$. The P-adic completion of $C\langle t_1 \ldots t_d \rangle$ is then the subring of all infinite sums $\Sigma a_k t^{[k]}$ such that a_k tends to zero P-adically as $|k| \to \infty$. Evidently this ring is $\mathbb{Z}$-torsion free. □

§4. Calculus with Divided Powers.

Suppose (S,I,γ) is a PD scheme (with I a quasi-coherent ideal, as always), and suppose X is an S-scheme. Let $X/S^{(\nu+1)}$ be the $\nu+1$-fold Cartesian product of X with itself, computed over S, and let $\Delta:X \to X/S^{(\nu+1)}$ be the diagonal immersion. The immersion Δ is locally closed and has $\nu+1$ retractions to X. It follows from Remark (3.20.6) that if γ extends to X, the divided power envelope of X in $X/S^{(\nu+1)}$ does not depend on γ.

4.1 Definition. Suppose γ extends to X and either $m\mathcal{O}_X = 0$ or X/S is separated. Then we can form the divided power envelope "$D_{X/S}(\nu)$" of X in $X/S^{(\nu+1)}$. The corresponding n^{th} order divided power neighborhood we shall denote by "$D^n_{X/S}(\nu)$" — note that it makes sense even without the hypotheses $m\mathcal{O}_X = 0$ or X/S separated.

As a consequence of Proposition (3.32), we see that if X/S is smooth, if $m\mathcal{O}_X = 0$, and if $x_1 \ldots x_n$ are local coordinates of X, then the structure sheaf $\mathcal{D}_{X/S}(1)$ of $D_{X/S}(1)$ is isomorphic to the PD polynomial algebra $\mathcal{O}_X\langle\xi_1 \ldots \xi_n\rangle$, where $\xi_i = (1\otimes x_i - x_i\otimes 1)$.

4.2 Remark. The natural map $P^1_{X/S} \xrightarrow{\alpha} \mathcal{D}^1_{X/S}(1)$ is an isomorphism. To see this, let $J \subseteq P_{X/S}$ be the ideal of X, so that $J/J^2 = \operatorname{Ker}[P^1_{X/S} \to \mathcal{O}_X]$. Since J/J^2 is a square zero ideal, (3.2.4) shows that is has a PD structure. Hence α has a section and therefore is injective. We know that $\mathcal{D}(1)$ is generated by $\{\xi^{[k]}: k \geq 1, \xi \in J\}$, and in $\mathcal{D}^1(1)$, $\xi^{[k]} = 0$ if $k > 1$. It follows from this that the map α is also surjective. For

$n > 1$, all we can say is that the map $D^n_{P^n_{X/S}}(X) \to D^n_{X \times_S X}(X)$ is an isomorphism.

We shall now indicate the PD version of stratifications. Recall that we had an algebra morphism:

$$\delta : \mathcal{O}_X \otimes_{\mathcal{O}_S} \mathcal{O}_X \longrightarrow (\mathcal{O}_X \otimes_{\mathcal{O}_S} \mathcal{O}_X) \otimes_{\mathcal{O}_X} (\mathcal{O}_X \otimes_{\mathcal{O}_S} \mathcal{O}_X) \text{ given by:}$$

$$x \otimes y \longmapsto x \otimes 1 \otimes 1 \otimes y \quad \text{, or}$$

$$\xi \longmapsto \xi \otimes 1 + 1 \otimes \xi \quad \text{if} \quad \xi = 1 \otimes x - x \otimes 1 \quad .$$

Since the augmentation ideal $\mathcal{J}$ of $\mathcal{D}_{X/S}(1)$ has a PD structure and is an augmentation ideal, (3.7) tells us that $\mathcal{D}_{X/S}(1) \otimes_{\mathcal{O}_X} \mathcal{D}_{X/S}(1)$ is a PD algebra. The universal mapping property of $\mathcal{D}$ tells us that δ induces a PD morphism $\mathcal{D}_{X/S}(1) \to \mathcal{D}_{X/S}(1) \otimes \mathcal{D}_{X/S}(1)$, which we shall again denote by δ. We have the useful formula: $\delta(\xi^{[k]}) = \delta(\xi)^{[k]} = (\xi \otimes 1 + 1 \otimes \xi)^{[k]} = \sum_{i+j=k} \xi^{[i]} \otimes \xi^{[j]}$, from which we see that δ induces maps $\mathcal{D}^{m+n}_{X/S}(1) \to \mathcal{D}^m_{X/S}(1) \otimes \mathcal{D}^n_{X/S}(1)$ for all m and n.

4.3 Definition. Let E be an $\mathcal{O}_X$-module. A "PD stratification on E" is a collection of isomorphisms:

$$\varepsilon_n : \mathcal{D}^n_{X/S}(1) \otimes E \to E \otimes \mathcal{D}^n_{X/S}(1)$$

such that:

1) Each ε_n is $\mathcal{D}^n_{X/S}(1)$-linear.

2) The ε_n's are compatible, in the obvious sense, and $\varepsilon_0 = \mathrm{id}_E$.

3) The following diagram commutes, for all m and n (cocycle condition):

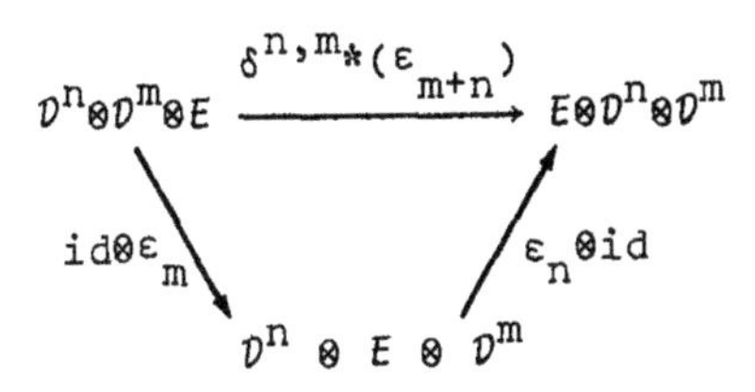

Somewhat stronger than a PD stratification is a "hyper PD stratification". (This notion is not useful in characteristic zero.)

4.3H <u>Definition.</u> An "HPD stratification on E" is an isomorphism: $\varepsilon:\mathcal{D}_{X/S}(1) \otimes E \to E \otimes \mathcal{D}_{X/S}(1)$ such that

1) ε is $\mathcal{D}_{X/S}$-linear.
2) ε reduces to the identity mod $\bar{J}$.
3) The cocycle condition holds. (Let the reader imagine the diagram.)

We can interpret PD stratifications in terms of an analogue of differential operators, called "PD differential operators". We have to be careful however: a PD differential operator $E\to F$ <u>cannot</u> be regarded as a map $E \to F$.

4.4 <u>Definition.</u> If E and F are $\mathcal{O}_X$-modules, a "PD differential operator $E \to F$ of order $\leq n$" is a $\mathcal{O}_X$-linear map $\mathcal{D}^n_{X/S}(1)\otimes E \to F$. An "HPD differential operator $E \to F$" is an $\mathcal{O}_X$-linear map $\mathcal{D}_{X/S}(1)\otimes E \to F$.

A PD differential operator $f:\mathcal{D}^n_{X/S}(1)\otimes E \to F$ induces a map $f^b: E \to F$, as in the diagram below. This f^b is a differential operator of order $\leq n$, but note that f^b does not necessarily determine f, if $n \geq 2$. This is because E does not generate $\mathcal{D}^n_{X/S}(1)\otimes E$ as on $\mathcal{O}_X$-module, as it did $P^n_{X/S}(1)\otimes E$.

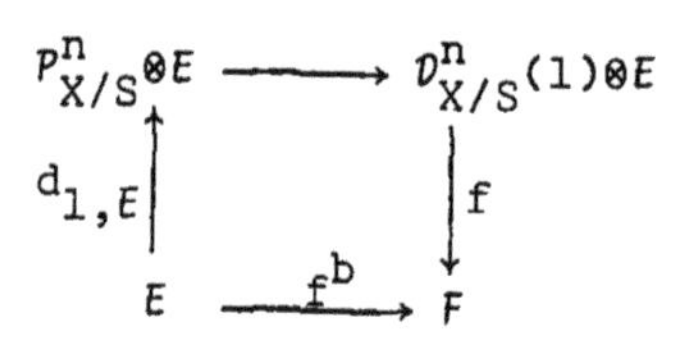

We are forced to define composition of PD differential operators formally: If $f:\mathcal{D}^n\otimes E \to F$ and $g:\mathcal{D}^m\otimes F \to G$, then $g\circ f$ is defined to be the composite: $g\circ(id_{\mathcal{D}^m}\otimes f)\circ(\delta^{n,m}\otimes id_E)$: $\mathcal{D}^{n+m}\otimes E \to G$. The diagram shows that $(g\circ f)^b = g^b\circ f^b$. The same definition works, of course, for HPD differential operators.

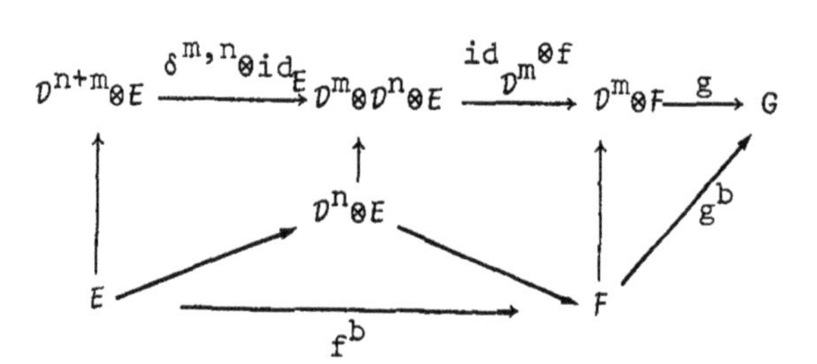

4.5 Example. Suppose $\partial:\mathcal{O}_X \to \mathcal{O}_X$ is a derivation. Then ∂ "is" a differential operator of order ≤ 1, hence also a PD differential operator of order ≤ 1 (c.f. the diagram)

$$\begin{array}{ccc} \mathcal{O}_X & \xrightarrow{\partial} & \mathcal{O}_X \\ \downarrow & & \uparrow D \\ P^1_{X/S} & \xrightarrow{\cong} & \mathcal{D}^1_{X/S} \end{array}$$

More precisely, $\partial = D^b$, where D is a unique PD differential operator of order ≤ 1. Suppose that $p\mathcal{O}_X = 0$. Then $\partial^p = \overbrace{\partial\circ\ldots\circ\partial}^{p \text{ times}}$ is again a derivation, hence a differential operator of order ≤ 1. On the other hand, $\overbrace{D\circ D\ldots\circ D}^{p \text{ times}} = D^p$ has order exactly p, in general, as we shall see. Of course, $(D^p)^b = \partial^p$, but D^p is <u>not</u> the P D differential operator $D^{(p)}$ of order ≤ 1 such that $(D^{(p)})^b = \partial^p$.

Suppose now that X/S is smooth and $(x_1\ldots x_m)$ is a system of local co-ordinates. Let $\xi_i = 1\otimes x_i - x_i\otimes 1$, and recall that $\{\xi_1^{[q]} = \xi_1^{[q_1]}\xi_2^{[q_2]}\ldots\xi_k^{[q_k]} : |q| \leq n\}$ is a basis for $\mathcal{D}^n_{X/S}(1)$ as a left $\mathcal{O}_X$-module. We want to describe composition in terms of the dual basis $\{D_q : |q| \leq n\}$ of $Hom_{\mathcal{O}_X}(\mathcal{D}^n_{X/S}(1),\mathcal{O}_X)$, the sheaf of P D differential operators of order $\leq n$.

4.6 <u>Proposition.</u> With the notations of the paragraph above, $D_q\circ D_{q'} = D_{q+q'}$.

<u>Proof.</u> We must compute $(D_q\circ D_{q'})(\xi^{[k]})$. By definition this is the image of $\xi^{[k]}$ under the composite:

$$\mathcal{D}^{2n} \xrightarrow{\delta^{n,n}} \mathcal{D}^n\otimes\mathcal{D}^n \xrightarrow{id\otimes D_{q'}} \mathcal{D}^n \xrightarrow{D_q} \mathcal{O}_X .$$

Now $\delta(\xi^{[k]}) = \sum_{i+j=k} \xi^{[i]}\otimes\xi^{[j]}$, which maps to $\xi^{[k-q']}$ under $id\otimes D_{q'}$ and finally to $\delta_{k,q+q'}$ under D_q. □

4.7 Corollary. If $D_i = D_{(0,0\ldots1\ldots0)} = \partial/\partial x_i$, $D_q = D_1^{q_1}D_2^{q_2}\ldots D_k^{q_k} = D^q$. □

Thus the ring of PD differential operators is generated by the first order ones — and composition is just "formal". It will follow that even in characteristic $p > 0$, a PD stratification on E is equivalent to an integrable connection on E.

4.8 Theorem. Suppose X/S is smooth and E is an $\mathcal{O}_X$-module. Then the following are equivalent:

i) A PD stratification $\{\varepsilon_n\}$ on E.

ii) A collection of $\mathcal{O}_X$-linear maps:

$$\text{PD } \mathcal{D}iff^n_{X/S}(\mathcal{O}_X,\mathcal{O}_X) \to \text{PD } \mathcal{D}iff^n_{X/S}(E,E)$$

which fit together to give a ring homomorphism:

$$\rho : \bigcup_0^\infty \text{ PD } \mathcal{D}iff^n_{X/S}(\mathcal{O}_X,\mathcal{O}_X) \longrightarrow \bigcup_0^\infty \text{ PD } \mathcal{D}iff^n_{X/S}(E,E)$$

ii bis) For all $\mathcal{O}_X$-modules F and G, maps

$$\rho_{F,G} : \text{PD } \mathcal{D}iff^n(F,G) \longrightarrow \text{PD } \mathcal{D}iff^n(E\otimes F, E\otimes G)$$

taking identities to identies and compatible with composition.

iii) An integrable connection ∇ on E.

Proof. (We have skipped the analogue of (2.11.2) to save space.) (i) ⇒ (ii): This is the same as in (2.11): Given a PD differential operator $u:\mathcal{D}^n \to \mathcal{O}_X$, we set

$$\rho_n(u) = \mathcal{D}^n \otimes E \xrightarrow{\varepsilon_n} E \otimes \mathcal{D}_n \xrightarrow{id_E \otimes u} E \quad .$$

(ii) ⇒ (i): The same method as in (2.11) will work as soon as we know that the maps ρ_n are automatically $\mathcal{D}^n$-linear. This requires a different argument. We need a formula which describes in local coordinates $x_1 \dots x_n$ the $\mathcal{D}^n$-module structure on

$$\text{PD } \mathcal{D}iff^n(\mathcal{O}_X, \mathcal{O}_X) \underset{\text{def.}}{=} Hom(\mathcal{D}^n, \mathcal{O}_X).$$

<u>Claim.</u> If $z \in \mathcal{D}_{X/S}(1)$, $zD^q = \sum_{i+j=q} ((i,j))D^i(z)D^j$, in the multi-index notation we have been using. To verify the claim, note that both sides of the equation are $\mathcal{O}_X$-linear maps $\mathcal{D}^{|q|}_{X/S}(1) \to \mathcal{O}_X$, so it suffices to show that they agree on any $y = \xi^{[s]}$ (s is again a multi-integer). Moreover, since both sides are also $\mathcal{O}_X$-linear in z, we may take $z = \xi^{[r]}$. Recalling that $(zD^q)(y)$ is by definition $D^q(zy)$, we get:

$$\begin{aligned} \xi^{[r]}D^q(\xi^{[s]}) &= D^q(\xi^{[r]}\xi^{[s]}) \\ &= D^q(((r,s))\xi^{[r+s]}) \\ &= ((r,s))\delta_{q,r+s}, \quad \text{while} \end{aligned}$$

$$\sum_{i+j=q} ((i,j))D^i(\xi^{[r]})D^j = ((r,q-r))D^{q-r} \quad , \quad \text{so}$$

$$\sum_{i+j=q} ((i,j))D^i(\xi^{[r]})D^j(\xi^{[s]}) = ((r,q-r))\delta_{q-r,s} = ((r,s))\delta_{q,r+s} \quad .$$

This proves the claim; we use it to check that ρ_n's are $\mathcal{D}^n$-linear, still working with local coordinates. We have to check that for any $z \in \mathcal{D}$, and any multi index q, $\rho(zD^q) = z\rho(D^q)$, i.e. $\rho(zD^q)(m) = \rho(D^q)(zm)$ for any $m \in \mathcal{D}\otimes E$

$$\text{Set } \theta_\alpha = \rho(D_\alpha), \text{ so that } \rho(D^q) = \theta^q \quad .$$

Since $\mathcal{D}^1 \cong P^1$, (2.11) tells us that ρ_1 is $\mathcal{D}^1$-linear, so that we at least have $z\theta_\alpha = \rho(zD_\alpha)$ for any z. By the above claim $zD_\alpha = D_\alpha(z) + \mathrm{id}(z)D_\alpha$, where $\mathrm{id} \in \mathrm{PD}\ \mathcal{D}iff(\mathcal{O}_X,\mathcal{O}_X)$ is the "identity" PD differential operator. (N.B. $\mathrm{id} \in \mathrm{PD}\ \mathcal{D}iff^n(F,F) = \mathcal{H}om_{\mathcal{O}_X}(\mathcal{D}^n \otimes F, F)$ is just the canonical projection.) Thus, we have:

$$\theta_\alpha(zm) = D_\alpha(z)m + \mathrm{id}(z)\theta_\alpha(m) \qquad \text{for any } z \text{ and } m \quad .$$

Now by induction it is easy to prove that

$$\theta^q(zm) = \sum_{i+j=q} ((i,j)) D^i(z)\,\theta^j(m) \qquad \text{for any multi-index } q, \text{ i.e.}$$

$$\rho(D^q)(zm) = \sum_{i+j=q} ((i,j))\, D^i(z)\rho(D^j)(m)$$

$$= \rho\Big[\sum_{i+j=q} ((i,j)) D^i(z) D^j\Big](m) = \rho(zD^q)(m) \ ,$$

as desired. It is interesting to note that the key in (2.11) was the fact that $d_1(\mathcal{O}_X)$ generates P ; here we use the fact that PD $\mathcal{D}iff^1$ generated PD $\mathcal{D}iff$.

(ii) ⇒ (iii) This is the same as before.

(iii) ⇒ (ii) Here we have to be careful in dealing with curvature, because of this distinction between u and u^b for PD differential operators. The problem is that in the formula we gave for curvature, only u^b's occur.

4.9 Lemma. Suppose ∇ is a connection on E and u and v are PD differential operators of order ≤ 1. Then the PD differential operator $\Delta = [\nabla(u),\nabla(v)]$ has order ≤ 1. (*Note:* smoothness of X/S is not needed here.)

Proof. First a warning: If we replace $\nabla(u)$ and $\nabla(v)$ by arbitrary operators of order ≤ 1 on E, the statement is false, even in characteristic 0, if rank $E > 1$.

To prove the lemma, we must show that Δ, the difference between the top and bottom compositions in the diagram below, factors through $\mathcal{D}^1_{X/S}\otimes E$ i.e. that it annihilates $K\otimes E$, where $K = \mathrm{Ker}(\mathcal{D}^2_{X/S} \to \mathcal{D}^1_{X/S})$

$$\mathcal{D}^2\otimes E \xrightarrow{\delta^{1,1}\otimes id} \mathcal{D}^1\otimes\mathcal{D}^1\otimes E \underset{id\otimes\nabla(u)}{\overset{id\otimes\nabla(v)}{\rightrightarrows}} \mathcal{D}^1\otimes E \underset{\nabla(v)}{\overset{\nabla(u)}{\rightrightarrows}} E$$

4.9.1 Claim. $\delta^{1,1}(K)$ is generated by $\{\xi\otimes\xi : \xi = 1\otimes x - x\otimes 1,\ x \in \mathcal{O}_X\}$. To check this, recall that $J = \mathrm{Ker}(\mathcal{O}_X\otimes\mathcal{O}_X \to \mathcal{O}_X)$ is generated by $\{\xi = 1\otimes x - x\otimes 1\}$ and $K = J^{[2]}/J^{[3]}$ is generated by $\{\xi^{[2]},\xi\xi'\}$. Now compute:

$$\delta^{1,1}(\xi^{[2]}) = \delta^{1,1}(\xi)^{[2]} = (\mathbb{1}\otimes\xi + \xi\otimes\mathbb{1})^{[2]} = \mathbf{1}\otimes\xi^{[2]} + \xi\otimes\xi + \xi^{[2]}\otimes\mathbf{1}$$

$= \xi\otimes\xi$ in $\mathcal{D}^1\otimes\mathcal{D}^1$. Next, $\delta^{1,1}(\xi\xi') = \delta^{1,1}(\xi)\delta^{1,1}(\xi')$

$= (\mathbb{1}\otimes\xi + \xi\otimes\mathbf{1})(\mathbb{1}\otimes\xi' + \xi'\otimes\mathbb{1}) = \xi'\otimes\xi + \xi\otimes\xi'$ (in $\mathcal{D}^1\otimes\mathcal{D}^1$)

$= (\xi+\xi')\otimes(\xi+\xi') - \xi\otimes\xi - \xi'\otimes\xi'$.

4.9.2 Claim. If $\xi \in \bar{J}/\bar{J}^{[2]} \subseteq \mathcal{D}^1$, if $u \in$ PD $\mathcal{D}iff^1(\mathcal{O}_X,\mathcal{O}_X)$, and if $m \in E$, then $\nabla(u)(\xi\otimes m) = u(\xi)m$.

Recall that $\nabla(u)$ is by definition the composite:
$\mathcal{D}^1 \otimes E \xrightarrow{\varepsilon} E \otimes \mathcal{D}^1 \xrightarrow{id \otimes u} E$. Since $\varepsilon \equiv id \mod \bar{J}$, $\varepsilon(1 \otimes m) \equiv m \otimes 1 \mod \bar{J}$, and since $\bar{J}^{[2]} = 0$ in $\mathcal{D}^1$, $\varepsilon(\xi \otimes m) = \xi\ \varepsilon(1 \otimes m) = \xi(m \otimes 1) = m \otimes \xi$.
Thus, $\nabla(u)(\xi \otimes m) = (id \otimes u)\ \varepsilon(\xi \otimes m) = (id \otimes u)(m \otimes \xi) = u(\xi)m$.

Now we can prove the lemma. Let us follow one of our generators of $\delta^{1,1}(K)$, tensored with an $m \in E$, along the top and bottom of our diagram. On the top: $\nabla(u) \circ (id \otimes \nabla(v))(\xi \otimes \xi \otimes m) = \nabla(u)(\xi \otimes v(\xi)m) = u(\xi)v(\xi)m$. Since this doesn't depend on the order of u and v, the same is true along the bottom, and the lemma is proved.

The fact that (iii) gives (ii) is now clear: In local co-ordinates, we have from the curvature formula that, since $K = 0$, $[\theta_i^b, \theta_j^b] = 0$, where $\theta_i = \nabla(D_i)$, and hence $[\theta_i, \theta_j]^b = 0$. But since $[\theta_i, \theta_j]$ has order ≤ 1, it follows that $[\theta_i, \theta_j] = 0$. Thus it makes sense to define $\rho : \text{PD}\ \mathcal{D}iff(\mathcal{O}_X, \mathcal{O}_X) \to \text{PD}\ \mathcal{D}iff(E, E)$ by $D^q \to \theta^q$ and to extend by $\mathcal{O}_X$-linearity. □

What conditions on an integrable connection correspond to a hyper-PD stratification? We shall answer this only for p-torsion schemes.

4.10 <u>Definition.</u> Suppose that $p^m \mathcal{O}_S = 0$, that X/S is smooth, and that $\{x_i\}$ is a set of local co-ordinates for X/S. A connection ∇ on an $\mathcal{O}_X$-module E is said to "quasi-nilpotent" (with respect to the co-ordinate system) iff for each open $U \subseteq X$ and all $s \in \Gamma(U, E)$, there exist an open covering $\{U_\alpha\}$ of U and integers $\{e_{i,\alpha}\}$, such that $[\nabla(\partial/\partial x_i)^b]^{e_{i,\alpha}}(s|_{U_\alpha}) = 0$ for all i.

4.11 Remark. We shall see that, if the connection is integrable, the condition for quasi-nilpotence is independent of the co-ordinate system. Note that in any case $(\partial/\partial x_i^b)^p = 0$ if $p\mathcal{O}_S = 0$, $(\partial/\partial x_i^b)^p \subseteq p\mathcal{O}_X$ in general, and so the operators $(\partial/\partial x_i)^b$ are nilpotent.

4.12 Theorem. Suppose $p^m\mathcal{O}_S = 0$, X/S is smooth, and E is an $\mathcal{O}_X$-module. Then the data of an HPD stratification on E is equivalent to the data of an integrable connection on E which is quasi-nilpotent.

Proof. Let $\varepsilon : D_{X/S} \otimes E \to E \otimes D_{X/S}$ be an HPD stratification on E, let $\rho : \text{HPD } \mathcal{D}iff(\mathcal{O}_X, \mathcal{O}_X) \longrightarrow \text{HPD } \mathcal{D}iff(E,E)$ be the corresponding map, and let $\theta : E \to E \otimes D_{X/S}$ be $\varepsilon \circ d_{1,E}$; i.e. $\theta(m) = \varepsilon(1 \otimes m)$. Thus if

$$D : D_{X/S} \longrightarrow \mathcal{O}_X \in \text{HPD } \mathcal{D}iff(\mathcal{O}_X, \mathcal{O}_X),$$

$$\rho(D) : D_{X/S} \otimes E \to E \quad \text{is} \quad (id_E \otimes D) \circ \varepsilon \quad \text{and}$$

$$\rho(D)^b : E \to E \quad \text{is} \quad (id_E \otimes D) \circ \theta .$$

In any local co-ordinate system $(x_1, \ldots x_n)$, we have $D_{X/S} \cong \mathcal{O}_X\langle \xi_1, \ldots, \xi_n \rangle$, with dual "basis" $D^q = (\partial/\partial x_1)^{q_1} \ldots (\partial/\partial x_n)^{q_n}$ for HPD $\mathcal{D}iff(\mathcal{O}_X, \mathcal{O}_X)$. For any section m of E, we can write $\theta(m)$ as a locally finite sum: $\theta(m) = \sum_q m_q \otimes \xi^{[q]}$, with $m_q \in E$. Of course, m_q is just $(id_E \otimes D^q)(\theta(m)) = \rho(D^q)^b(m)$. Thus our formula is:

$$\theta(m) = \sum_q \rho(D^q)^b(m) \otimes \xi^{[q]} .$$

Since the sum is locally finite, $\rho(D^q)^b(m) = 0$ for almost all q, so the connection is quasi-nilpotent (in any co-ordinate system), as claimed.

Conversely, suppose ∇ is an integrable connection, quasi-nilpotent in some co-ordinate system. The connection ∇ defines $\rho(D_i)^b$ for each i; because of the rule for composition, we can make sense of $\rho(D^q)^b$ as $\prod_i [\rho(D_i)^b]^{q_i}$. Then quasi-nilpotent tells us that $\rho(D^q)^b(m) = 0$ for almost all q, so that we can use the displayed formula above to define $\theta: E \to E \otimes D_{X/S}$. This map is $\mathcal{O}_X$-linear (using the $\mathcal{O}_X$-structure from $D_{X/S}$ on the tensor product); we get a $D_{X/S}$-linear map ε $D_{X/S} \otimes E \to E \otimes D_{X/S}$ by extension of scalars. The only thing we must check carefully is the cocycle condition, paying attention to the distinction between D and D^b. We need the commutativity of the following diagram:

$$\begin{array}{ccc} E & \xrightarrow{\theta} & E\otimes\mathcal{D} \\ \theta\downarrow & & \downarrow\theta\otimes\mathrm{id} \\ E\otimes\mathcal{D} & \xrightarrow{\mathrm{id}\otimes\delta} & E\otimes\mathcal{D}\otimes\mathcal{D} \end{array}$$

This presents no difficulty:

$$(\theta\otimes\mathrm{id})(\theta(m)) = (\theta\otimes\mathrm{id})(\sum_q \rho(D^q)^b(m)\otimes\xi^{[q]}) = \sum_q\sum_{q'} \rho(D^{q'})^b\rho(D^q)^b(m)\otimes\xi^{[q']}\otimes\xi^{[q]}$$

$$= \sum_{q,q'} \rho(D^{q'+q})^b(m)\otimes\xi^{[q]}\otimes\xi^{[q']}$$

$$= \sum_n \rho(D^n)^b(m)\otimes \sum_{q+q'=n} \xi^{[q]}\otimes\xi^{[q']} = \sum_n \rho(D^n)^b(m)\otimes\delta(\xi)^{[n]}$$

$$= (\mathrm{id}\otimes\delta)\cdot\theta(m). \quad \square$$

4.13 <u>Corollary.</u> The condition of quasi-nilpotence of an integrable connection is independent of the co-ordinate system.

4.14 <u>Exercise.</u> A connection is quasi-nilpotent iff its reduction mod p is.

§5. The Crystalline Topos.

We are ready to assemble the constructions of the first four chapters into the notion of the "crystalline site", which will then give rise to the "crystalline topos". In this and the next section, all schemes will be killed by a power of a prime p, unless otherwise specified. This assumption will allow us to postpone the technical difficulties of inverse limits.

Let $S = (S, I, \gamma)$ be a PD-scheme, which will play the role of the "base". For any S-scheme X to which γ extends (in the sense of (3.14)), we want to define the "crystalline site of X relative to S", which we denote by Cris(X/S). It is the site whose objects are pairs $(U \hookrightarrow T, \delta)$, where U is a Zariski open subset of X, $U \hookrightarrow T$ is a closed S-immersion defined by an ideal J, and δ is a PD structure on J which is compatible with γ, in the sense of (3.17). Note that since $\mathcal{O}_T$ is killed by a power of p, J is a nil ideal, so that $U \to T$ is a homeomorphism. We shall often abuse notation by writing (U, T, δ) for $(U \hookrightarrow T, \delta)$, or even by just writing T for the whole thing. We shall call $T = (U \hookrightarrow T, \delta)$ an "S-PD thickening of U". The assumption that γ extends to X insures us that for each Zariski open U of X, $(U \xrightarrow{id_U} U, 0)$ is an object of Cris(X/S), because then γ is compatible with the trivial PD-structure 0 on the zero ideal of U. (The converse is also true – in fact, the reader can easily check that if U has <u>any</u> S-PD thickening at all, γ extends to U.)

We must also specify the morphisms of the site Cris(X/S) and the covering families. A morphism $T \xrightarrow{u} T'$ in Cris(X/S) is just a commutative square:

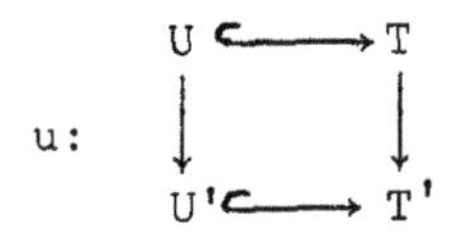

such that the arrow $U \to U'$ is an inclusion in the Zariski topology of X and the arrow $T \to T'$ is an S-PD morphism $(T,J,\delta) \to (T',J',\delta')$. A covering family of Cris(X/S) is a collection of morphisms $\{u_i : T_i \to T\}$ such that each $T_i \to T$ is an open immersion and $T = \cup T_i$. [Note that the PD structures δ_i are determined by δ, and conversely.] It is immediate that these satisfy the necessary conditions for covering families (or pretopologies in the language of [SGA4 II No. 1]).

A standard situation for the reader to keep in mind is $S = \mathit{Spec}\ W(k)/p^n W(k)$, $I = (p)$, with the canonical PD structure γ, and X a scheme over *Spec* k. (In particular, X will usually not be smooth over S.) A more general situation which arises naturally is Cris(Y/T), where Y is an X-scheme and $X \hookrightarrow T$ is a PD thickening.

Grothendieck's philosophy is that the notion of a site is too rigid and a better structure is the "topos" to which it gives rise, namely the category of sheaves of sets on the site. Different sites can give rise to equivalent topoi, but the "geometry" (e.g. the cohomology) of all such sites should be the same. It is possible to give an intrinsic description of what makes a category a topos [SGA 4, IV, 1], in keeping with this philosophy, but we shall not do so here.

We denote the topos of sheaves on Cris(X/S) by $(X/S)_{cris}$. It is the full subcategory of the category of presheaves (cofunctors) Cris(X/S) $\rightsquigarrow$ ((Sets)) whose objects satisfy the sheaf axiom: For every covering family $\{T_i \to T\}$, the sequence of sets:

$$0 \to F(T) \to \prod_i F(T_i) \rightrightarrows \prod_{i,j} F(T_i \cap T_j) \quad \text{is exact.}$$

We can give a somewhat more explicit description of an object F of $(X/S)_{cris}$ as follows: Fix, for the moment, some object (U,T,δ) of Cris(X/S). For each Zariski open $T' \hookrightarrow T$, let $U' = U \cap T'$ and let $\delta' = \delta|_{T'}$; then there is a morphism $(U',T',\delta') \to (U,T,\delta)$ in Cris(X/S): and hence a map of sets $F(U,T,\delta) \to F(U',T',\delta')$. It is clear that the assignment $T' \to F(U',T',\delta')$ defines, in fact, a sheaf $F_{(U,T,\delta)}$ on the Zariski topology of T. If $u:(U_1,T_1,\delta_1) \to (U,T,\delta)$ is a morphism in Cris(X/S), then for any open set $T' \subseteq T$ let $T_1' = u^{-1}(T')$, so that we have a morphism: $(U_1',T_1',\delta_1') \to (U',T',\delta')$ and hence a map $F(U',T',\delta') \to F(U_1',T_1',\delta_1')$, i.e. $F_T(T') \to F_{T_1}(T_1')$. In other words, the map u induces a map of sheaves $F_T \to u_* F_{T_1}$, or, equivalently, a map $u^{-1}F_T \to F_{T_1}$. The reader can easily convince himself that this construction establishes the following:

5.1 Proposition. *A sheaf F on Cris(X/S) is equivalent to the following data:*

For every S-PD thickening (U,T,δ) in Cris(X/S), a Zariski sheaf F_T on T, and for every morphism $u:(U_1,T_1,\delta_1) \to (U,T,\delta)$

in Cris(X/S), a map $\rho_u : u^{-1}(F_T) \to F_{T_1}$, satisfying the following compatibilities:

i) If $v: (U_2,T_2,\delta_2) \longrightarrow (U_1,T_1,\delta_1)$ is another morphism, then the diagram:

$$\begin{array}{ccccc} v^{-1}u^{-1}F_T & \xrightarrow{v^{-1}(\rho_u)} & u^{-1}F_{T_1} & \xrightarrow{\rho_v} & F_{T_2} \\ \| & & & & \| \\ (u\circ v)^{-1}(F_T) & & \xrightarrow{\rho_{u\circ v}} & & F_{T_2} \end{array}$$

commutes.

ii) If $u: T_1 \to T$ is an open immersion, the map

$\rho_u^{-1} : u^{-1}(F_T) \to F_{T_1}$ is an isomorphism.

5.2 Examples. 1) The cofunctor $T \mapsto \mathcal{O}_T$ defines a sheaf of rings on Cris(X/S), which we call the "structure sheaf" and denote by $\mathcal{O}_{X/S}$. This is the most important object of our study.

2) The cofunctor $T \mapsto \mathcal{O}_U$ defines another sheaf of rings on Cris(X/S), which we denote by $\mathcal{O}_X$ or (in a notation to be developed later) $i_{X/S*}\mathcal{O}_X$.

3) The cofunctor $T \mapsto \mathrm{Ker}\{\mathcal{O}_T \to \mathcal{O}_U\}$ defines a sheaf of PD ideals in $\mathcal{O}_{X/S}$, which we denote by $\mathcal{J}_{X/S}$. It is also extremely important. We have an exact sequence:

$$0 \to \mathcal{J}_{X/S} \to \mathcal{O}_{X/S} \to i_{X/S*}(\mathcal{O}_X) \to 0 \quad .$$

5.3 Remark. A useful consequence of the Zariski interpretation (5.1) of $(X/S)_{cris}$ is the fact that it has enough points. For us, this means that we can tell if a map of sheaves: $\nu: F \to G$ is an isomorphism by looking at stalks: It is enough to check that for each $x \in X$ and each S-PD thickening T of a *Zariski neighborhood of* x, *the map of stalks:* $(F_T)_x \to (G_T)_x$ is an isomorphism.

In order to exploit Grothendieck's philosophy we recall the natural embedding of a (suitable) site into its associated topos. *If* T *is an object in any category* X, $\tilde{T} = \mathcal{H}om[\ ,T]$ is an object in the category $\hat{X}$ of presheaves on X, and for any $F \in \hat{X}$, there is a canonical identification à la Yoneda: $\mathcal{H}om_X[\tilde{T},F] \cong F(T)$. For most sites, and certainly for Cris(X/S) (as the reader will easily verify), $\tilde{T}$ *is in fact* a sheaf, so that one has a fully faithful functor from the site into its associated topos. Incidentally, we shall often find it useful to abuse notation a bit and write $F(G)$ for $\mathcal{H}om_X[G,F]$ if F and G are presheaves, even if G is not representable by an object of the site.

The first major advantage of the crystalline topos over the crystalline site is its functoriality. If $g: X' \to X$ is an S-morphism, there is no way to pull back S-PD thickenings in X to S-PD thickenings in X', in general. However, we will be able to pull back the sheaves they represent, and hence obtain a morphism of topoi $(X'/S)_{cris} \to (X/S)_{cris}$. It is wise to recall what this means:

5.4 Definition. A morphism of topoi $f\colon \mathcal{T}' \to \mathcal{T}$ is a functor: $f_*\colon \mathcal{T}' \to \mathcal{T}$ which has a left adjoint $f^*\colon \mathcal{T} \to \mathcal{T}'$ which commutes with finite inverse limits.[1]

5.5 Remark. Of course, f_* and f^* determine each other uniquely. Moreover f_* commutes with arbitrary inverse limits and f^* with arbitrary direct limits just from the adjointness. The intuition behind the extra condition on f^* is that the "stalks" of f^*F are supposed to be the same as the stalks of F, and finite inverse limits can be computed stalk by stalk. It is remarkable that this condition is sufficient to give f geometric meaning. If there is danger of confusion with module pull-back, we sometimes write f^{-1} for f^*.

Suppose that in the commutative diagram below, $S' \to S$ is a PD morphism. Then we want to obtain a morphism of topoi $g_{cris}\colon (X'/S')_{cris} \to (X/S)_{cris}$. We begin by specifying $g^*_{cris}(\tilde{T})$ if $T \in Cris(X/S)$.

(5.6.1)
$$\begin{array}{ccc} X' & \xrightarrow{g} & X \\ \downarrow & & \downarrow \\ S' & \xrightarrow{f} & S \end{array}$$

5.6 Definition. Suppose $(U,T,\delta) \in Cris(X/S)$. Then $g^*(U,T,\delta)$ is the sheaf on $Cris(X'/S')$ defined by

$$(U',T',\delta') \longmapsto \text{PD } \mathcal{H}om_g\,[(U',T',\delta'),(U,T,\delta)]\ .$$

This means the following: $g^*(T)(T')$ is the empty set unless $g(U') \subseteq U$, and if $g(U') \subseteq U$, it is the set of all PD maps $h\colon (T',\delta') \to (T,\delta)$ making the diagram below commute:

[1] This definition differs slightly from the usual one, c.f. [SGA4 IV 3.1.1 and 3.2.2].

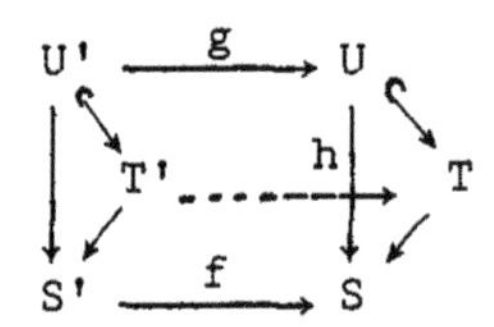

It is easy to see that this presheaf is in fact a sheaf — which may or may not be representable.

It turns out that there is a unique way to make a morphism of topoi $g_{cris}: (X'/S')_{cris} \to (X/S)_{cris}$ compatible with the above definition, i.e., such that $g^*_{cris}(\tilde{T}) = g^*(T)$ for every $T \in Cris(X/S)$. The uniqueness is just abstract nonsense, but the existence requires some algebraic geometry. Let's do the abstract part first.

5.7 Proposition. Let C' and C be categories, with corresponding presheaf categories $\hat{C}'$ and $\hat{C}$. Suppose $\varphi: C \to \hat{C}'$ is a functor. Then there is a unique pair of functors $\varphi^{\cdot}: \hat{C} \to \hat{C}'$ and $\varphi_{\cdot}: \hat{C}' \to \hat{C}$ such that $\varphi^{\cdot}|_C = \varphi$ and such that $\varphi^{\cdot}$ is left adjoint to $\varphi_{\cdot}$.

Proof. First we establish the uniqueness of (and the formula for) $\varphi_{\cdot}: \hat{C}' \to \hat{C}$. If $G' \in \hat{C}'$, then $\varphi_{\cdot}(G')$ is a cofunctor from C to ((Sets)), and if $T \in C$, $\varphi_{\cdot}(G')(T) = Hom_{\hat{C}}[\tilde{T}, \varphi_{\cdot}(G')] = Hom_{\hat{C}'}[\varphi^{\cdot}(\tilde{T}), G')] = Hom_{\hat{C}'}[\varphi(T), G']$. This proves the existence and uniqueness of $\varphi_{\cdot}$:

(5.7.1) $\varphi_{\cdot}(G')(T) = Hom_{\hat{C}'}[\varphi(T), G']$, if $T \in C$ and $G' \in \hat{C}'$.

Since $\varphi_.$ is unique, so is its adjoint $\varphi^{\cdot}$; we have only to show that it exists. Again, we'll give a formula for it. Fix some $T' \in C'$, and consider the category $\varphi\{T'\}$ of all morphisms $\{\tilde{T}' \to \varphi(T): T \in C\}$. Then:

(5.7.2) $\varphi^{\cdot}(G)(T') = \varinjlim_{\varphi\{T'\}} G(T)$, if $G \in \hat{C}$ and $T' \in C'$.

This is easy to verify if we aren't too careful about the compatibilities: We have to check that if $F' \in \hat{C}'$, $Hom_{\hat{C}'}[\varphi^{\cdot}(G),F'] \overset{\cong}{\leftrightarrow} Hom_{\hat{C}}[G,\varphi_.(F')]$. Now a morphism: $\varphi^{\cdot}(G) \xrightarrow{h} F'$ means a compatible collection of maps:

$$h_{T'}: \varphi^{\cdot}(G)(T') \to F'(T') \qquad \forall\ T' \in C' .$$

Since $\varphi^{\cdot}(G)(T') = \varinjlim\{G(T): u \in \varphi(T)(T')\}$, to give h is the same as giving a compatible collection of set maps: $G(T) \to F'(T')$, indexed by triples (T,T',u) where $u \in \varphi(T)(T')$. This is equivalent to a family of maps $G(T) \times \varphi(T)(T') \to F'(T')$, or equivalently, $G(T) \to Hom_{sets}[\varphi(T)(T'), F'(T')]$. Now the compatibility in T' amounts to saying that we in fact have a morphism of sets: $G(T) \to Hom_{\hat{C}'}[\varphi(T), F'] = \varphi_.(F')(T)$, and compatibility in T means that we have a morphism of cofunctors: $G \to \varphi_.(F')$. □

We now apply this to the crystalline site.

5.8 Proposition. Suppose we have a square as in (5.6.1), and we let φ: Cris(X/S) → Cris(X'/S')^ be defined by g^*, as in (5.6). Then:

(5.8.1) For any $T \in \mathrm{Cris}(X/S)$, $\varphi(T) = g^*(T)$ is a sheaf on $\mathrm{Cris}(X'/S')$.

(5.8.2) For any sheaf G' on $\mathrm{Cris}(X'/S')$, $g_{\mathrm{cris}_*}(G') \underset{\mathrm{def}}{=} \varphi_\cdot(G')$ (as in (5.7.1)) is a <u>sheaf</u> on $\mathrm{Cris}(X/S)$.

(5.8.3) For any sheaf G on $\mathrm{Cris}(X/S)$, let $g_{\mathrm{cris}}{}^*(G)$ be the sheafification of the presheaf $\varphi^\cdot(G)$ (5.7.2). Then $g'_{\mathrm{cris}}{}^*$ is left adjoint to g_{cris_*}.

<u>Proof.</u> We have already asserted (5.8.1). Let us check (5.8.2), which almost purely abstract nonsense. Suppose $\{T_i \to T\}$ is a covering family in $\mathrm{Cris}(X/S)$ and $\{s_i \in g_*(G)(T_i)\}$ is a compatible family. We must show that there is a unique section $s \in g_*(G)(T)$ inducing $\{s_i\}$. By definition, s is to be a morphism of sheaves $g^*(T) \to G$. Thus, for each $T' \in \mathrm{Cris}(X'/S')$, we must *find a map of sets:* $s_{T'}: g^*(T)(T') \to G(T')$, *compatible* with change of T'. So suppose $h \in g^*(T)(T')$, i.e., h is an S-PD morphism $T' \to T$. Let $T'_i = h^{-1}(T_i)$, so that we have a *covering* $\{T'_i \to T'\}$ *of* T', and maps $h_i: T'_i \to T_i \in g^*(T_i)(T'_i)$ induced by h. Now the given s_i are sheaf maps $g^*(T_i) \to G$, and hence we have $s_{iT'_i}: g^*(T_i)(T'_i) \to G(T'_i)$. Evaluating at h_i, we *obtain elements* $s_{iT'_i}(h_i)$ *of* $G(T'_i)$. *The sheaf axiom for* G and the compatibility of the s_i's insure that there is a unique element $s_{T'}(h)$ of $G(T')$ inducing $\{s_{iT'_i}(h_i)\}$. It is easy to *see that the set maps* $s_{T'}$ *we have thus defined are compatible* with change in T', and hence that s is indeed a morphism

$g^*(T) \to G$, as desired. Uniqueness is also clear. It remains only to verify (5.8.3). But this is easy: If G is a sheaf on Cris(X/S) and G' a sheaf on Cris(X'/S'), we have by the universal property of sheafification and by (5.7):

$$\mathit{Hom}[g_{cris}{}^*(G),G'] = \mathit{Hom}[\varphi^{\cdot}(G),G'] = \mathit{Hom}[G,\varphi_{\cdot}(G')] =$$

$$= \mathit{Hom}[G,g_{cris_*}(G')] \ . \quad \square$$

To know that we have constructed a morphism of topoi, we have to see that g^*_{cris} preserves finite inverse limits. This is not abstract nonsense, and rests on crucial properties of PD structures.

5.9 Proposition. $g_{cris}{}^*$ commutes with finite inverse limits.

Proof. Let J be a finite category and $\{G_j : j \in J\}$ an inverse system in $(X/S)_{cris}$. There is a natural map of sheaves: $g_{cris}{}^*(\varprojlim G_j) \to \varprojlim g_{cris}{}^*(G_j)$; to show it is an isomorphism it is enough to check the stalks, as explained in (5.3). If G is a sheaf on Cris(X/S), if (U',T',δ') is an object of Cris(X'/S'), and if $x' \in T'$, then the stalk of $g^*(G)$ at (T',x') is $\varinjlim\{G(T) \mid h\colon V' \to T\}$. Here we are taking the limit over the category $I_{x',T',g}$, whose objects are S-PD maps $h\colon V' \to T$, with V' an open neighborhood of x' in T' and T an S-PD thickening of an open set U of X. Since we are taking a direct limit, we should think of an arrow $h_1 \to h_2$ as a diagram:

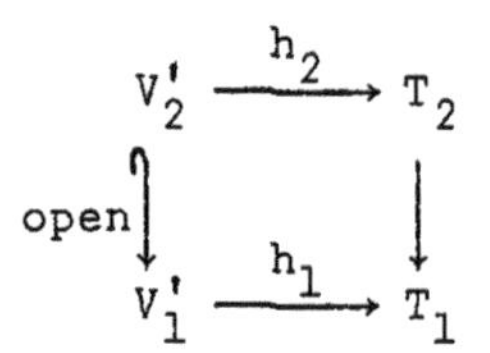

Thus, our problem is to show that the map

$\varinjlim_I \varprojlim_J G_j(T) \longrightarrow \varprojlim_J \varinjlim_I G_j(T)$ is an isomorphism, where

$I = I_{x',T',g'}$. Now it is well-known that finite inverse limits commute with filtering direct limits. Thus, the following suffices:

5.10 Proposition. The category $I = I_{x',T',g}$ is filtering. In other words:

i) It's not empty.

ii) It's connected, i.e. given objects a and b, there exists an object c and maps $a \to c$ and $b \to c$.

iii) Given maps $a \to b$ and $a \to c$, there exists an object d and a commutative diagram:

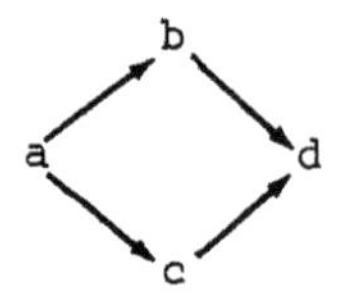

iv) If $u: a \to b$ and $v: a \to b$ are two maps, then there exists a map $w: b \to c$ such that $w \circ u = w \circ v$.

Proof. The properties above follow from three divided power constructions, the first of which goes back to Grothendieck's letter to Tate.

5.11 Lemma. Suppose (U',T',δ') is an object of $\mathrm{Cris}(X'/S')$ with U' affine and $g(U')$ contained in an open affine U of X. Then there exists an S-PD thickening $U \hookrightarrow T$ and an S'-PD morphism $T' \to T$ extending the map $g: U' \to U$.

Proof. We may assume S and S' are affine, given by (A,I,γ) and (A',I',γ'), respectively, with $U' = \mathit{Spec}\ C'$, $T' = \mathit{Spec}\ B'$, and $U = \mathit{Spec}\ C$. Let $B = C \times_{C'} B'$ in the following diagram:

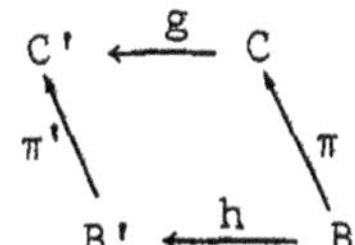

Since π' is surjective, so is π ; its kernel is the ideal $J = \{(b',0) \in B' \times_{C'} C : b' \in J'\}$, where J' is the kernel of π'. Clearly we can set $\delta_n(b',0) = (\delta_n' b',0)$ to get a PD structure on J and making h a PD morphism. We must check that δ is compatible with γ . First, note that γ extends to C and γ' to B', by hypothesis, and since $A \to A'$ is a PD morphism, γ also extends to B'. Thus we can extend γ to B by setting $\bar{\gamma}_n(b',c) = (\gamma_n(b'), \gamma_n(c))$ if $(b',c) \in IB$, i.e. if $b' \in IB'$ and $c \in IC$. Since δ' is compatible with γ', it is compatible with γ , so δ is also compatible with γ . □

The above lemma shows that the category $I_{X,T',g}$ is not empty; the following lemma shows that it satisfies conditions (ii) and (iii).

5.12 Lemma. *Suppose we are given $T' \in \mathrm{Cris}(X'/S')$, T_1 and $T_2 \in \mathrm{Cris}(X/S)$, an S-scheme Y, and finally the solid arrows in the diagram below. Then there exists a $T \in \mathrm{Cris}(X/S)$ and dotted arrows making the diagram commute.*

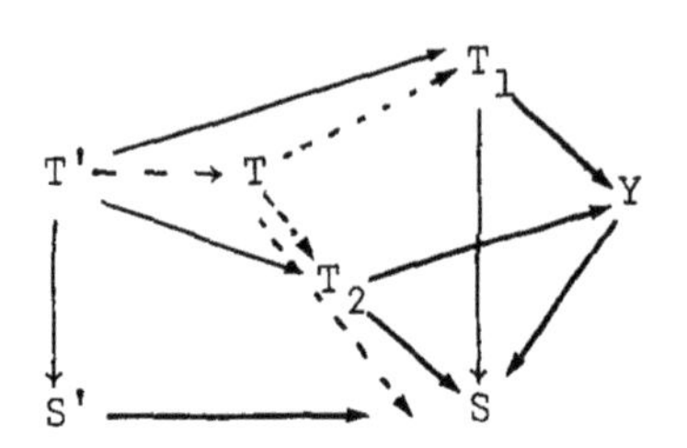

Proof. Let $U = U_1 \cap U_2$ in X, which maps to the fiber product $T_1 \underset{Y}{\times} T_2$ by a locally closed immersion. We take T to be the divided power envelope $D_{U,\delta_1,\delta_2}(T_1 \times T_2)$ of U in $T_1 \underset{Y}{\times} T_2$. The subscript δ_1,δ_2 means that we impose compatibility with the two PD structures δ_1 and δ_2 on T_1 and T_2, in a manner analogous to the construction of $D_{U,Y}$. For details, see [Berthelot III, 2.1.3]. □

5.13 Lemma. *Suppose $T' \in \mathrm{Cris}(X'/S')$, $T_1 \in \mathrm{Cris}(X/S)$, Y and Z are schemes, and we are given the solid arrows in the diagram. Assume that $u_1 \circ h = u_2 \circ h$ and that $u_1|_{U_1} = u_2|_{U_1}$. Then there exists a $T \in \mathrm{Cris}(X/S)$ and dotted arrows shown, such that $u_1 \circ u = u_2 \circ u$.*

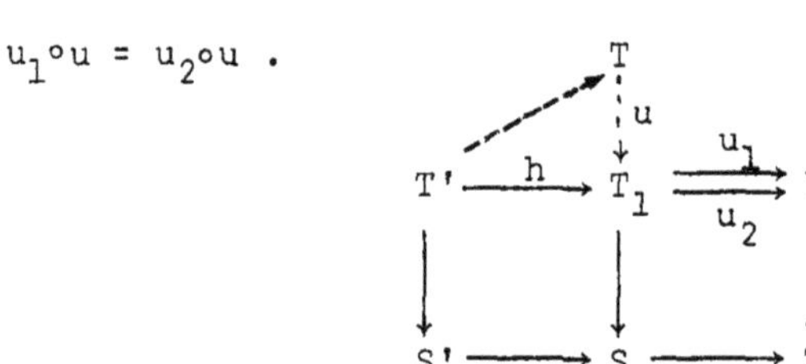

<u>Proof.</u> Let J_1 be the PD ideal $\operatorname{Ker}(\mathcal{O}_{T_1} \to \mathcal{O}_{U_1})$. Since u_1 and u_2 agree on U_1, $(u_1^* - u_2^*)(\mathcal{O}_Y) \subseteq J_1$. Let J be the sub PD ideal of J_1 generated by $(u_1^* - u_2^*)(\mathcal{O}_Y)$, and let T be the subscheme of T_1 it defines. □

These lemmas complete the proof of Proposition (5.10) — (in order to get condition (iv) of Proposition (5.10), one takes Z = S and Y some object of Cris(X/S) in Lemma (5.13)). The general statement of the lemma is needed when one has a morphism $g'\colon X'' \to X'$ over $S'' \to S'$, and one wants an identification $g_{cris} \circ g'_{cris} \cong (g \circ g')_{cris}$. The trick is to show that that $g'^*(g^*(T)) \cong (g \circ g')^*(T)$ for every T of Cris(X/S). We shall leave this as an exercise for the assiduous reader. □

5.14 <u>Remark.</u> The construction of Lemma 5.12 tells us, in essence, that finite products are representable in Cris(X/S). If we combine this with the Lemma 5.13 (and the universal mapping property of the construction not stated there) we conclude that inverse limits over finite nonempty index sets are representable in Cris(X/S). For details, cf. [Berthelot, III, 2.1].

The structure sheaf $\mathcal{O}_{X/S}$ is a sheaf of rings in $(X/S)_{cris}$, and we shall want to consider $(X/S_{cris}, \mathcal{O}_{X/S})$ systematically as a ringed topos. If $g\colon X' \to X$ is as in (5.6.1), then $g_{cris}\colon (X'/S')_{cris} \to (X/S)_{cris}$ is a morphism of ringed topoi, i.e. there is a natural map $\mathcal{O}_{X/S} \to g_{cris_*}\mathcal{O}_{X'/S'}$: if T is an object of Cris(X/S), we need a map

$\mathcal{O}_{X/S}(T) \to (g_{cris_*}\mathcal{O}_{X'/S'})(T) = \mathcal{H}om(g^*(T), \mathcal{O}_{X'/S'})$. But if T is an object of $\mathrm{Cris}(X'/S')$ and $h \in g^*(T)(T')$, h is a map $T' \to T$ and hence provides us with a map of rings: $\mathcal{O}_{X/S}(T) = \mathcal{O}_T(T) \to \mathcal{O}_{T'}(T') = \mathcal{O}_{X'/S'}(T')$.

We are now in a good position to use general nonsense about topoi to define cohomology. First, the notion of global sections.

5.15 Definition. Let X be a site with topos $\mathcal{T}$ and let T be an object of X. Then "$\Gamma(T, \ )$" is the functor $\mathcal{T} \to ((\mathrm{Sets}))$ given by $F \mapsto F(T)$. More generally, if T is any object of $\mathcal{T}$, $\Gamma(T, \ )$ is the functor $F \mapsto \mathcal{H}om_{\mathcal{T}}(T,F)$. If e is the final of $\mathcal{T}$, we write $\Gamma(\mathcal{T},F)$ or $\Gamma(F)$ for $\Gamma(e,F)$.

The final object e of a topos $\mathcal{T}$ is just the sheafification of the constant presheaf whose value at any U is the set {0} consisting of a single element. In the case of the category of sheaves on an ordinary topological space X, this sheaf is represented by the open set X of the site, but in the case of the crystalline topos, it is not representible. In general, a section $s \in \Gamma(\mathcal{T},F) = \mathcal{H}om(e,F)$ is just a compatible collection of sections $s_T \in F(T)$ for every object T of X, i.e. an element of $\varprojlim_{T \in X} F(T)$.

If A is any sheaf of rings on a site X, the category of sheaves of A-modules has enough injectives, and for any $T \in \mathcal{T}$ we can define $H^i(T, \ )$ to be the ith derived functor of $\Gamma(T, \ )$, in this category. As usual, the abelian group structure of $H^i(T,F)$ does not depend on A, i.e., can be computed in the category of abelian sheaves on X. Moreover, one has a Leray spectral sequence for any morphism of topoi. As a consequence:

5.16 Proposition. Suppose $g: X' \to X$ covers a PD morphism $S' \to S$. Then g induces a morphism of topoi: $g_{cris}: (X'/S')_{cris} \to (X/S)_{cris}$. If E' is an abelian sheaf in $(X'/S')_{cris}$, there is a Leray spectral sequence:
$E_2^{pq} = H^p(X/S_{cris}, R^q g_{cris_*} E') \Rightarrow H^i(X'/S'_{cris}, E')$.

Proof. Let us sketch the construction of the Leray spectral sequence of a morphism of topoi $f: \mathcal{T}' \to \mathcal{T}$. The key is the fact that f^* preserves finite inverse limits. In particular, it takes the final object of $\mathcal{T}$ to the final object of $\mathcal{T}'$, because the final object is the inverse limit over the empty category. Then if E' is a sheaf in $\mathcal{T}'$, $\Gamma(\mathcal{T}',E') \cong \Gamma(\mathcal{T},f_*E')$, because $\Gamma(\mathcal{T},f_*E') = \mathcal{H}om_{\mathcal{T}}[e,f_*E'] = \mathcal{H}om_{\mathcal{T}'}[f^*e,E'] = \mathcal{H}om_{\mathcal{T}'}[e',E'] = \Gamma(\mathcal{T}',E')$. We can therefore apply the spectral sequence of a composite functor, if f_* takes injectives to injectives. But since its left adjoint f^* is exact, this is automatic. □

We can already prove the first important rigidity property of crystalline cohomology, namely its invariance under certain PD thickenings:

5.17 Theorem. Suppose we have a Cartesian square as shown, with $S_0 \hookrightarrow S$ defined by a sub PD ideal K of I. Then there is a natural isomorphism: $H^i(X/S_{cris}, \mathcal{O}_{X/S}) \to H^i(X_0/S_{cris}, \mathcal{O}_{X_0/S})$.

$$\begin{array}{ccc} X_0 & \xrightarrow{i} & X \\ \downarrow & & \downarrow \\ S_0 & \hookrightarrow & (S,I,\gamma) \end{array}$$

Proof. We have a morphism of topoi $i_{cris}: (X_0/S)_{cris} \to (X/S)_{cris}$, and hence a corresponding Leray spectral sequence. It is therefore clear that the last two of the following statements will imply the theorem:

5.17.1 If $T \in \mathrm{Cris}(X/S)$, $i^*(T)$ is representable in $\mathrm{Cris}(X'/S')$.

5.17.2 The functor i_{cris_*} is exact.

5.17.3 $i_{cris_*}(\mathcal{O}_{X_0/S}) \cong \mathcal{O}_{X/S}$.

Indeed, let U be the open set of X defined by the PD ideal (J,δ) of $\mathcal{O}_T$, and let $U_0 = U \cap X_0$. Then $U_0 \hookrightarrow U$ is defined by the ideal $K\mathcal{O}_U$ of $\mathcal{O}_U$, and hence $U_0 \hookrightarrow T$ is defined by the ideal $K\mathcal{O}_T + J$. By definition of the crystalline site, γ extends to a PD structure $\bar{\gamma}$ on $I\mathcal{O}_T$. It is easy to see that the fact that K is a sub PD ideal of I implies that $K\mathcal{O}_T$ becomes a sub PD ideal of $I\mathcal{O}_T$. Since $\bar{\gamma}$ is compatible with δ, $K\mathcal{O}_T + J$ has a PD structure compatible with γ, by (3.16). Thus (U_0,T) becomes an object of $\mathrm{Cris}(X_0/S)$, and it is clear that the morphism $(U_0,T) \to (U,T)$ is universal, i.e. that (U_0,T) represents $i^*(U,T) = \mathcal{H}om(\ \ ,(U,T))$.

It is now easy to compute $i_{cris_*}(F)$ for any sheaf F on $\mathrm{Cris}(X_0/S)$. By definition, $i_{cris_*}(F)(U,T) = \mathcal{H}om[i^*(U,T),F] =$ $= \mathcal{H}om[(U_0,T),F] = F(U_0,T)$. In terms of the associated Zariski sheaves, we have therefore: $i_{cris_*}(F)_{(U,T)} \cong F_{(U_0,T)}$. Now since a sequence of sheaves is exact iff the associated sequence of Zariski sheaves for every PD thickening is, it is clear that i_{cris_*} is exact. Also, $i_{cris_*}(\mathcal{O}_{X_0/S})(U,T) = \mathcal{O}_{X_0/S}(U_0,T) =$ $= \mathcal{O}_T(T) = \mathcal{O}_{X/S}(U,T)$, so $i_{cris_*}(\mathcal{O}_{X_0/S}) = \mathcal{O}_{X/S}$. □

Remark. We shall see later (6.2) that i_{cris_*} is exact whenever i is a closed S-immersion.

We shall now describe some constructions which will help us to localize certain calculations on the crystalline site. The first of these is a projection from the crystalline topos to the Zariski topos X_{zar}. This projection will be a fancy way of fitting together the crystalline cohomology of the various Zariski open subsets of X.

5.18 Proposition. There is a natural morphism of topoi: $u_{X/S}\colon (X/S)_{cris} \to X_{zar}$, given by:

(1) For $F \in (X/S)_{cris}$ and $j\colon U \hookrightarrow X$ an open immersion,

$(u_{X/S_*}(F))(U) = \Gamma(U/S_{cris}, j^*_{cris} F)$

(2) For $E \in X_{zar}$ and $(U,T,\delta) \in \mathrm{Cris}(X/S)$,

$(u_{X/S}{}^*(E))(U,T,\delta) = E(U)$.

Notice that $u_{X/S_*}(F)(U)$ is the set of global sections of F over $(U/S)_{cris}$, (which should be thought of as "horizontal" sections over U). It is quite easy to see that u_{X/S_*} and $u_{X/S}{}^*$ are adjoint to each other. Moreover, $(u_{X/S})^*(E)_{(U,T,\delta)} \cong E|_U$ for any (U,T,δ), and from this it is clear that $u_{X/S}{}^*$ commutes with arbitrary inverse limits. Thus we really do have a morphism of topoi. It is not however, a morphism of ringed topoi, because there is, in general, no map $\mathcal{O}_X \to u_{X/S_*}\mathcal{O}_{X/S}$. The Leray spectral sequence of $u_{X/S}$ nevertheless exists. (In terms of de Rham cohomology, it is, in fact, just the so-called conjugate spectral sequence: $E_2^{pq} = H^p(X, \mathcal{H}^q(\Omega^{\cdot}_{X/S})) \Rightarrow \mathbb{H}^i(X, \Omega^{\cdot}_{X/S})$, if X/S is smooth. This will become apparent later.)

5.19 To justify calling $u_{X/S}$ a projection we provide it with a section $i_{X/S}: X_{zar} \to (X/S)_{cris}$. The functor $u^*_{X/S}$ itself has a left adjoint, which deserves to be called $u_{X/S!}$, given by $u_{X/S!}(F) = F_X$; i.e. the Zariski sheaf given by F on the object $(X,X,0)$ of Cris(X/S). Clearly $\mathcal{H}om_{zar}[u_{X/S!}F,E] = \mathcal{H}om_{cris}[F,u^*_{X/S}E]$. Since $u_{X/S!}$ clearly commutes with inverse limits, we get our morphism of topoi $i_{X/S}$ by setting $i_{X/S}{}^* = u_{X/S!}$ and $i_{X/S_*} = u_{X/S}{}^*$. Obviously, $u_{X/S_*} \circ i_{X/S_*} = id$. Unlike $u_{X/S}$, $i_{X/S}$ is a morphism of ringed topoi, because there is a map $\mathcal{O}_{X/S} \to i_{X/S_*}\mathcal{O}_X \cong \mathcal{O}_X$. Since i_{X/S_*} is obviously exact, the Leray spectral sequence of $i_{X/S}$ is degenerate, and we have:

5.20 $H^*(X_{zar}, E) \cong H^*(X/S_{cris}, i_{X/S_*}E)$ for any Zariski sheaf E on X .

5.21 There is another, more general, notion of localization we shall need, which makes sense in any topos. In fact, the definition works in any category, we shall give only a sketch, for more details, compatibilities, and points of view, the reader can look in [SGA 4 IV 8 and SGA 3 I].

5.22 Proposition. If T is an object in a category C, let $C|_T$ denote the category of arrows in C with target T, and let $s: C|_T \to C$ denote the functor which takes an arrow $f: S \to T$ into its source S .

5.22.1 If $X \times T$ exists in C for all X, then s has a right adjoint $r_T: C \to C|_T$, given by $X \mapsto [pr_2: X \times T \to T]$.

5.22.2 If $\hat{}$ means the corresponding presheaf category, there is a natural equivalence of categories:

$$\hat{C}|_{\hat{T}} \longleftrightarrow \widehat{C|_T} \ .$$

5.22.3 If $Z \in \hat{C}$, then the functor $r_Z: \hat{C} \to \hat{C}|_Z$ has a right adjoint j_Z .

Proof. The first statement is obvious. For the next one, let us content ourselves with a description of $\eta: \hat{C}|_{\hat{T}} \to \widehat{C|_T}$ and $\zeta: \widehat{C|_T} \to \hat{C}|_{\hat{T}}$; the reader who so desires can easily check that they are quasi-inverses. If $F: Z \to \hat{T}$ is an object of $\hat{C}|_{\hat{T}}$, then $\eta(F)$ is the cofunctor $C|_T \to ((Sets))$ given by the following: If $f: S \to T \in C|_T$, then $f \in \hat{T}(S)$, and we just take $\eta(F)(f) = F_S^{-1}(f)$, where $F_S: Z(S) \to \hat{T}(S)$ is the function induced by F. Going back, if G is an object of $\widehat{C|_T}$,

Let $S_G: \hat{C} \to ((\text{Sets}))$ by $S_G(X) = \prod_{f \in \hat{T}(X)} G(f)$; then $\zeta(G): S_G \to \hat{T}$ is the obvious map. Note that a morphism $G: S \to T$ induces an object $\hat{G}_1: \hat{S} \to \hat{T}$ of $\hat{C}|_{\hat{T}}$ and an object $G_2 = \mathrm{Hom}_{\hat{C}|_{\hat{T}}}[\ \ ,G]$ of $\hat{C}|_{\hat{T}}$, and it is immediate that $\hat{G}_2 = \eta(\hat{G}_1)$. We shall make these identifications without further reference.

Before explaining (5.22.3), let us remark that $\hat{C}$ has products, hence r_Z exists (and is compatible with r_T, if $Z = \hat{T}$ and r_T exists).

Using this construction one gets internal *Hom* in $\hat{C}$: If Z' and Z are objects of $\hat{C}$,

5.22.4 $$\mathit{Hom}[Z',Z](T) = \mathit{Hom}_{\hat{C}|_{\hat{T}}}[r_{\hat{T}}(Z'), r_{\hat{T}}(Z)] .$$

Now we can describe the functor j_Z:

5.22.5 If $F: Z' \to Z$ is an object of $\hat{C}|_Z$, $j_Z(F)$ is the presheaf of cross-sections of F.

Let us verify that for any $G \in \hat{C}$, $\mathit{Hom}[r_Z(G), F] = \mathit{Hom}[G, j_Z(F)]$. An arrow $r_Z(G) \to F$ is a commutative triangle:

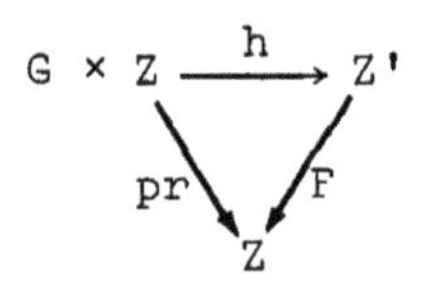

This means a compatible collection of triangles, for each $T \in \mathrm{Ob}(C)$,

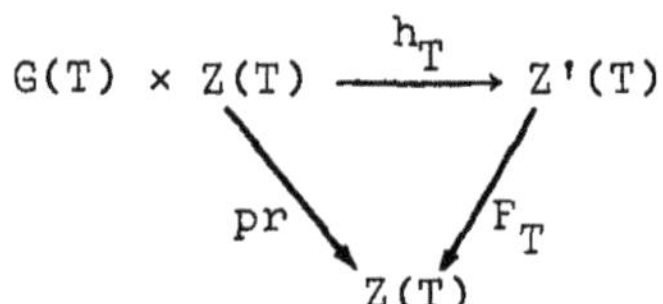

These triangles are clearly equivalent to the data: for each T and each $g \in G(T)$ a cross-section of $F_T\colon Z'(T) \to Z(T)$... i.e. a morphism of cofunctors: $G \longrightarrow j_Z(F)$. □

Now if C has a topology, one can perform the same constructions in the category $\tilde{C}$ of sheaves on C, and if the functor $C \to \hat{C}$ factors through $\tilde{C}$, all the above constructions go through without change. One obtains:

5.23 <u>Proposition.</u> Let T be a topos, Z an object of T. Then $T|_Z$ is also a topos, and there is a morphism of topoi:

$$j_Z\colon T|_Z \to T$$

in which j_{Z*} is given as in (5.22.4), and $j_Z{}^* = r_Z$ (5.22.1). Moreover, $j_Z{}^*$ has a left adjoint $j_{Z!}$ given by the functor s in (5.22.1).

<u>Proof.</u> Let's just observe that since $j_Z{}^*$ has a left adjoint, it preserves arbitrary inverse limits. □

5.24 <u>Proposition.</u> If E is an abelian sheaf in T there is a canonical isomorphism

$$H^i(T|_Z\, , j_Z{}^*(E)) \xrightarrow{\cong} H^i(Z,E)\ .$$

<u>Proof.</u> The final object of $T|_Z$ is $id: Z \to Z$, and $j_{Z!}(id_Z)$ is just Z. Thus, if $E \in T$, we have
$\Gamma(Z,E) = Hom_T(Z,E) = Hom_T[j_{Z!}(id_Z),E] = Hom_{T|_Z}[id_Z\, , j_Z{}^*(E)] =$
$= \Gamma(T|_Z\, , j_Z{}^*(E))$. This proves the statement for $i = 0$; the

general case will follow from the facts that j_Z^* is <u>exact</u> and takes injectives to injectives. The exactness of j_Z^* (true for any morphism of topoi) follows in this case from the existence of its right adjoint j_{Z_*} and its left adjoint $j_{Z!}$ (which is not the same for abelian sheaves as it is for sheaves of sets). The statement about injectives will be a consequence of the fact that $j_{Z!}$ is also exact, as is clear from its description: If $F \in T|_X$ is an abelian sheaf, then $j_{Z!}(F)$ is the sheaf associated to the presheaf $Y \to \bigoplus_{u \in [Y,Z]} F(u)$. □

We can make the localization construction somewhat more concrete in the case of the crystalline topos by giving a description in terms of compatible families of associated Zariski sheaves, in analogy with (5.1). We omit the proof.

5.25 <u>Proposition.</u> If $Z = (U,Z,\varepsilon)$ is an object of Cris(X/S), then $(X/S)_{cris|\tilde{Z}}$ may be described as follows: It is (equivalent to) the category of compatible collections of pairs (u,F_u), where $u\colon Y \to Z$ is a morphism in Cris(X/S) and F_u is a Zariski sheaf on Y. Moreover, if $j_Z\colon (X/S)_{cris|\tilde{Z}} \to (X/S)_{cris}$ is the canonical morphism, and if we also use the description (5.1) of $(X/S)_{cris}$, then j_{Z*} and $j_Z{}^*$ are given by:

5.25.1 If $E \in (X/S)_{cris}$ and $u\colon Y \to Z \in \text{Cris}(X/S)|_Z$, then $j_Z{}^*(E)_u$ is the Zariski sheaf E_Y on Y .

5.25.2 If $F = \{(u, F_u)\} \in (X/S)_{cris|\tilde{Z}}$ and $T = (U,T,\delta) \in Cris(X/S)$, then $j_{Z*}(F)_T$ is the Zariski sheaf $p_{T*}(F_{p_Z})$ on T, where p_T and p_Z are the PD-morphisms:

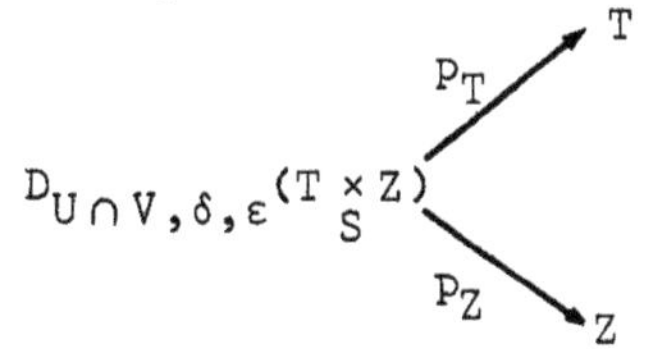

(Here D is the double PD-envelope used in (5.12).)

<u>Proof.</u> If $f: G \to \tilde{Z}$ is an object of $(X/S)_{cris|\tilde{Z}}$, here is how to obtain the corresponding collection $\{(u, F_u)\}$: If $u: Y \to Z$ is a morphism in Cris(X/S), and if $Y' \hookrightarrow Y$ is a Zariski open subset, then $F_u(Y') = Hom_{\tilde{Z}}(\tilde{Y}', G)$. We let the reader verify that this gives a sheaf on Y_{zar}, and that we have an equivalence of categories.

Now it is easy to verify (5.25.1), using the description (5.22.1) of j_Z^*. Since $j_Z^*(E)_u(Y)$ is, by definition, $Hom_Z(\tilde{Y}, j_Z^*(E))$, and since $j_Z^*(E)$ is $pr_Z: E \times \tilde{Z} \to \tilde{Z}$, this is just $Hom(\tilde{Y}, E) = E_Y(Y)$. Localizing on the Zariski topology of Y gives (5.25.1).

The final statement is only a trifle more complicated. If $T' \hookrightarrow T$ is a Zariski open subset,

$j_{Z*}(F)_T(T') = Hom[T', j_{Z*}(F)] = Hom_{\tilde{Z}}[j_Z^*(\tilde{T}'), F] =$
$= Hom_{\tilde{Z}}(pr_Z: \tilde{T}' \times \tilde{Z} \to \tilde{Z}, F)$. But the "sheaf" $pr_{\tilde{Z}}: \tilde{T}' \times \tilde{Z} \to \tilde{Z}$ is represented by the "object" $p_Z: D_{U \cap V,\varepsilon,\delta}(T' \times Z) \to Z$ of $Cris(X/S)|_Z$, and hence by the description of F_u, $Hom_{\tilde{Z}}(pr_Z, F)$ is $F_{p_Z}(D(T' \times Z))$. Now formation of $D(T' \times Z)$ is compatible with localization in the Zariski topology of T, so (5.25.2) follows. □

We can relate our two notions of localization by an important diagram. This diagram will be our main computational device in the next two chapters.

5.26 Proposition. *Suppose* $Z = (V,Z,\varepsilon)$ *is an object of* Cris(X/S). *Then there is a commutative diagram of topoi:*

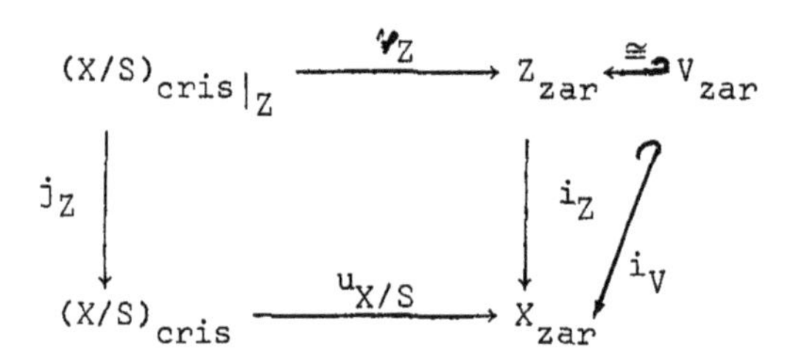

Moreover:

5.26.1 If E is a sheaf in $(X/S)_{cris}$, $\varphi_{Z_*}(j_Z^*(E)) \cong E_Z$.

5.26.2 φ_{Z*} is exact.

5.26.3 If E is an abelian sheaf in $(X/S)_{cris}$, $H^i(Z,E) \cong H^i(Z_{zar}, E_Z)$.

Proof. The morphism φ_Z is defined as follows: If $F = \{(u,F_u): u \in \text{Cris}(X/S)|_Z\}$ is an element of $(X/S)_{cris}|_Z$, $\varphi_{Z_*}(F)$ is just the Zariski sheaf F_{id_Z}. If E is a sheaf on Z_{zar} , $\varphi_Z^*(E) = \{(u,u^*(E)): u: T \to Z\}$. The reader can easily check that these form an adjoint pair, and that φ_Z^* preserves finite inverse limits. To see that the diagram commutes, it suffices to check that $j_Z^*u_{X/S}^*E \cong \varphi_Z^* i_Z^*E$ for every Zariski sheaf E on X. But $u_{X/S}^*(E)_{(U,T,\delta)} = E_U$, so $j_Z^*u_{X/S}^*(E)_u$

is just E_U, if $u: (U,T,\delta) \to (V,Z,\varepsilon)$. But $\varphi_Z^* \, i_Z^*(E) = u^*(i_Z^*(E)) = E_U$ also.

Now suppose E is a sheaf in $(X/S)_{cris}$. We recall from (5.25.1) that $j_Z^*(E)_u = E_T$, if $u: T \to Z$, hence $\varphi_{Z*} j_Z^*(E) = j_Z^*(E)_{id_Z} = E_Z$, proving (5.26.1). A morphism $F \to G$ in $(X/S)_{cris}|_Z$ is an epimorphism iff each morphism of Zariski sheaves $F_u \to G_u$ is, hence φ_{Z*} is exact. It follows that the Leray spectral sequence for the morphism φ_Z degenerates, so that if E is an abelian sheaf in $(X/S)_{cris}$, we have from (5.24):

$$H^i(Z,E) \cong H^i(X/S_{cris}|_Z, j_Z^*(E)) \cong H^i(Z_{zar}, \varphi_{Z*} j_Z^* E)$$

$$\cong H^i(Z_{zar}, E_Z). \quad \square$$

5.27 <u>Corollary.</u> Suppose that in the localization diagram (5.26) above, $V = X$, i.e., Z is an S-PD thickening of X. Then:

5.27.1 The functor j_{Z_*} is exact.

5.27.2 If E is an abelian sheaf in $(X/S)_{cris}|_Z$, $j_{Z_*}E$ is acyclic for u_{X/S_*}.

<u>Proof.</u> Look at the description (5.25.2) of the functor j_{Z_*}. If $T = (U,T,\delta) \in \text{Cris}(X/S)$, $U \subseteq V$, so that $D = D_{U,\delta,\varepsilon}(T \times_S Z)$ is a thickening of U. But then the map $p_T: D \to T$ is a homeomorphism, so p_{T_*} is exact, and (5.27.1) follows.

Now if E is an abelian sheaf in $(X/S)_{\mathrm{cris}|Z}$, we have a spectral sequence: $E_2^{pq} = R^p u_{X/S_*} R^q j_{Z_*}(E) \Rightarrow R^i(u_{X/S} \circ j_Z)_*(E)$, which degenerates to an isomorphism:
$R^i u_{X/S_*}(j_{Z_*}E) \cong R^i(u_{X/S} \circ j_Z)_*(E)$. But $u_{X/S} \circ j_Z = i_X \circ \varphi_Z$, and i_{X_*} and φ_{Z_*} are exact, so $R^i(u_{X/S} \circ j_Z)_*(E) = 0$ if $i > 0$. □

5.28 Let us conclude this section with a discussion of coverings of the final object e of $(X/S)_{\mathrm{cris}}$. Although e is not representable in Cris(X/S), it can often be covered by a representable sheaf. In fact, I claim that this is the case if there is a closed immersion $i: X \hookrightarrow Y$, with Y/S smooth. Then it is easy to see that $\tilde{Y} = i^*_{\mathrm{cris}}(Y)$ is represented by the S-PD envelope $X \hookrightarrow D_{X,Y}(Y)$ of X in Y, and I claim that the map $\tilde{Y} \to e$ is an epimorphism. This just means that the associated map of Zariski sheaves $\tilde{Y}_T \to e_T$ is an epimorphism, i.e., that for sufficiently small open subsets T' of T, $\tilde{Y}(T')$ is not empty. This in turn just means that the map $T' \cap X \to Y$ can be lifted to T', which follows from the fact that Y/S is smooth and $T' \cap X \hookrightarrow T'$ is a nilimmersion.

(N.B. It is a standard fact from SGA1 that a morphism with a smooth target Y locally extends over a nilpotent immersion; here we need it for nilimmersions as well. To see that this is also true, suppose $T' = \mathit{Spec}\ B$, with $X \cap T'$ defined by the nil ideal J. To see that the map $\tilde{Y}(B) \to \tilde{Y}(B/J)$ is surjective, write $J = \cup J_\lambda$, with $J_\lambda \subseteq J$ a finitely generated ideal. Then $B/J = \varinjlim B/J_\lambda$, and since Y/S is locally of finite presentation, $\tilde{Y}(B/J) = \varinjlim \tilde{Y}(B/J_\lambda)$ [EGA IV, vol. 3]. But each J_λ is a nil ideal and finitely generated, hence a nilpotent ideal, hence each $\tilde{Y}(B) \to \tilde{Y}(B/J_\lambda)$ is surjective, and what we need is a consequence of exactness of direct limits.)

According to SGA 4V (4.4 and 5.2), the above means that $\tilde{Y} \to e$ is a covering in $(X/S)_{cris}$. For us, this will mean that one can verify exactness by pulling back to $\tilde{Y}$. This is easy to see at least in the crystalline case: If $\Sigma = E_1 \to E_2 \to E_3$ is a 0-sequence of abelian sheaves on $(X/S)_{cris}$, and if $j_{\tilde{Y}}^*(\Sigma)$ is exact, then I claim that for each $T \in \text{Cris}(X/S)$ $\Sigma_T = E_{1,T} \to E_{2,T} \to E_{3,T}$ (and hence also Σ) is exact. Since the claim is local on T, we may assume that there is a morphism $u: T \to \tilde{Y}$, and by assumption the sequence $j_{\tilde{Y}}^*(\Sigma)_u$ is exact. But according to (5.25.1), this is just Σ_T.

5.29 The above remarks furnish the basis of the Cech-Alexander technique of computing cohomology in a topos. Although we shall not need this technique for the main comparison theorems, let us sketch it. If $\tilde{Y} \to e$ is a covering of the final object in a topos $\mathcal{T}$ (i.e., an epimorphism) consider the semi-simplicial object in $\mathcal{T}$; $\mathbb{N} \to \mathcal{T}: \nu \mapsto \tilde{Y}^{\nu+1}$, with the usual projection maps:

$$\cdots \longrightarrow \tilde{Y} \times \tilde{Y} \times \tilde{Y} \overrightarrow{\overrightarrow{\longrightarrow}} \tilde{Y} \times \tilde{Y} \rightrightarrows \tilde{Y} \quad .$$

If E is an abelian sheaf in $\mathcal{T}$, we can form a complex of sheaves $C^{\bullet}_{\tilde{Y}}(E)$ in $\mathcal{T}$, with $C^{\nu}_{\tilde{Y}}(E) = j_{\tilde{Y}^{\nu+1}*} j_{\tilde{Y}^{\nu+1}}^*(E)$ and with coboundary maps induced by the alternating sum of the projection maps.

If T is an object of $\mathcal{T}$, we have by (5.23) that $C^{\nu}_{\tilde{Y}}(E)(T) = \mathcal{H}om[j_{\tilde{Y}^{\nu+1}}{}^*(T), E] = E(T \times \tilde{Y}^{\nu+1})$. The value of this construction is in the fact that there is a natural resolution:

$$E \to C^{\bullet}_{\tilde{Y}}(E).$$

The map is clear, and we may verify that it is a quasi-isomorphism after restricting to $\mathcal{T}|_{\tilde{Y}}$. In other words, it suffices to check that the complex:

$$E(T) \to E(T \times \tilde{Y}) \to E(T \times \tilde{Y} \times \tilde{Y}) \to \cdots$$

is acyclic if T maps to $\tilde{Y}$. But in this case it is easy to construct a chain homotopy, using the maps $T \times \tilde{Y} \to T \times Y^{\nu+1}$ induced from the given $T \to \tilde{Y}$. The reader can work out the indices.

§6. Crystals.

A "crystal", says Grothendieck, is characterized by two properties: it is <u>rigid</u>, and it <u>grows</u>. Any sheaf F on Cris(X/S) "grows" over PD thickenings (U,T,δ) of open subsets of X by construction. In order for F to be a crystal, we impose a rigidity which we shall only make precise for sheaves of $\mathcal{O}_{X/S}$-modules. From now on, we shall write u^{-1} for pull-back of a sheaves of sets, and u^* for module pull-back (i.e. u^{-1} followed by tensor product with the structure sheaf) – unless there is no danger of confusion.

6.1 <u>Definition.</u> A "crystal" of $\mathcal{O}_{X/S}$-modules is a sheaf F of $\mathcal{O}_{X/S}$-modules such that for any morphism $u:(U',T',\delta') \to (U,T,\delta)$ in Cris(X/S), the map $u^*F_{(U,T,\delta)} \to F_{(U',T',\delta')}$ is an isomorphism.

A trivial example of a crystal is the sheaf $\mathcal{O}_{X/S}$ itself. An extremely nontrivial and useful example is furnished by the following:

6.2 <u>Proposition.</u> *Suppose* $i\colon Y \hookrightarrow X$ *is a closed immersion of* (S,I,γ)*-schemes (to which* γ *extends, as always). Then the functor* i_{cris_*} *is exact, and* $i_{cris_*}(\mathcal{O}_{Y/S})$ *is a crystal of* $\mathcal{O}_{X/S}$*-algebras.*

<u>Proof.</u> The key to this result is a close look at the sheaf $i^{-1}(T) = i^*(T)$ (5.6), if $T = (U,T,\delta) \in \text{Cris}(X/S)$. First of all, I claim that it is representable. Indeed,

$U_0 = U \cap Y$ is a closed subscheme of T, and it is natural to expect that $i^*(T)$ is represented by a suitable PD envelope of U_0 in T, compatible with δ as well as with the PD structure γ of the base. This is easily constructed. Suppose $U \hookrightarrow T$ is defined by the PD ideal (J, δ); then by compatibility of γ and δ, δ extends to a PD structure $\bar{\delta}$ on $J_1 = IO_T + J$, and we have:

6.2.1 Lemma. With the notations of the previous paragraph, $i^*(T)$ is represented by $D = (U_0, D_{U_0,\bar{\delta}}(T), [\ ]) \in \mathrm{Cris}(Y/S)$. If F is a sheaf in $(Y/S)_{\mathrm{cris}}$, then the Zariski sheaf associated to $i_{\mathrm{cris}_*}(F)$ on T is given by

$$i_{\mathrm{cris}_*}(F)_T = \lambda_*(F_D) \quad ,$$

where $\lambda\colon D \to T$ is the natural map.

Proof of 6.2.1. First we should check that D really is a PD thickening of U_0. Recall from (3.20.4) that for this we need to know that $\bar{\delta}$ extends to O_{U_0}. Since the image of $J_1 = IO_T + J$ in O_{U_0} is just IO_{U_0}, to which γ is assumed to extend, this is clear, as is the compatibility of $[\ ]$ with γ. Thus D really is an object of Cris(Y/S). We obtain the map λ from the universal mapping property of D, and it is easy to see that it makes D represent $i^*(T)$. If F is a sheaf on Cris(Y/S), we can now easily compute $i_{\mathrm{cris}_*}(F)$, as follows: By definition, $i_{\mathrm{cris}_*}(F)(T) = \mathcal{H}om(i^*(T), F) = F(D)$. In terms of associated Zariski sheaves, this gives the statement of the lemma. □

Proof of 6.2. The exactness of i_{cris_*} follows from the lemma, because as far as underlying Zariski spaces go, $\lambda\colon D \to T$ is the closed immersion $U_0 \to U$, so λ_* is exact.

It is clear from the above construction that to prove that $i_{cris_*}(\mathcal{O}_{Y/S})$ is a crystal, we have to establish a compatibility of the divided power envelope construction with base change. Namely, if $u\colon (U',T',\delta') \to (U,T,\delta)$ is a morphism in Cris(X/S), we have to show that the map $u^*(i_{cris_*}(\mathcal{O}_{Y/S})_T) \to i_{cris_*}(\mathcal{O}_{Y/S})_{T'}$, i.e. $\mathcal{O}_{T'}\otimes\mathcal{D}(T) \to \mathcal{D}(T')$, is an isomorphism.

This fact is one of the key technicalities in the theory of crystals, and to prove it we shall have to look again at the construction of PD envelopes. We may assume that $U = U'$ and that all our schemes are affine, with $T = \mathit{Spec}\ B$, $T' = \mathit{Spec}\ B'$, and $u\colon (B,J,\delta) \to (B',J',\delta')$ a PD morphism. Since $U = U'$, the map $B/J \to B'/J'$ is an isomorphism. Let $L \subseteq B$ and $L' \subseteq B'$ define the closed subscheme $U_0 = U \cap Y$. Then we have a commutative diagram as shown, and the theorem amounts to establishing the claim below:

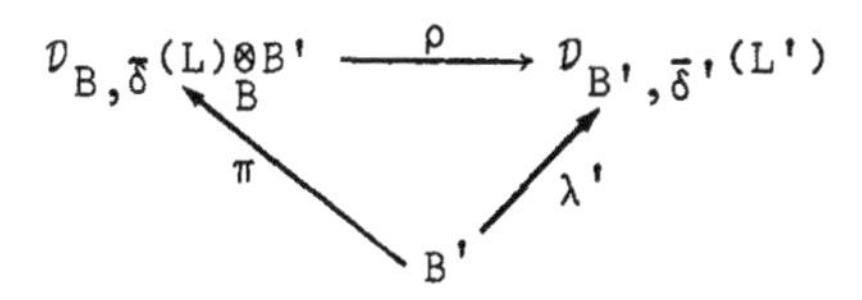

Claim. The arrow ρ above is an isomorphism.

Recall that since $L \not\supseteq J_1 = IB + J$, we construct $(\mathcal{D},\bar{L},[\ ])$ by setting $L_1 = L + J_1$, forming $(\mathcal{D},\bar{L}_1,[\ ])$, and taking $\bar{L}$ to be the sub-PD ideal of $\bar{L}_1$ generated by L. Now in our case $B/J \cong B'/J'$ and L' and LB' have the same image in B'/J', so $L' = LB' + J'$. It follows that $\pi(L')$ is contained in the image of $\bar{L}\otimes B'\oplus \mathcal{D}\otimes J'$ in $\mathcal{D}\otimes B'$, and hence also in the image L_1'' of $\bar{L}_1\otimes B'\oplus \mathcal{D}\otimes J_1'$. I claim that L_1'' has a PD structure compatible with $\bar{\delta}'$. The universal mapping property of $\mathcal{D}$ will then provide us with an inverse to the arrow ρ.

To get the desired PD structure on L_1'' we have to go back to the construction of $\mathcal{D}_{B,\bar{\delta}}(L_1)$ as $\Gamma_B(L_1)/K$, where K is a certain ideal of $\Gamma_B(L_1)$, c.f. (3.19). Now $\Gamma_B(L_1)\otimes_B B' \cong \Gamma_{B'}(L_1\otimes_B B')$, by (A2), which in degree zero is just B', and in particular contains the PD ideal $(J_1',\bar{\delta}')$. Because the PD ideal $\Gamma_B^+(L_1)\otimes B' \cong \Gamma_{B'}^+(L_1\otimes B')$ is an augmentation ideal, its PD structure [ ] is compatible with $\bar{\delta}'$, by (3.20.6). That is, the ideal $L_1^* = J_1'\Gamma_B(L) + \Gamma_B^+(L_1)\otimes B'$ has a PD structure δ^*, compatible with [ ] and $\bar{\delta}'$. The image of L_1^* in $\mathcal{D}_{B,\bar{\delta}}(L_1)\otimes B'$ is our ideal L_1'' , so to endow the latter with the compatible PD structure, we have to prove:

Claim. $K'\cap L_1^*$ is a sub PD ideal of L_1^* , where K' is the kernel of $\Gamma_B(L_1)\otimes B' \to \mathcal{D}_{B,\bar{\delta}}(L_1)\otimes B'$, or equivalently, the image of $K\otimes B'$ in $\Gamma\otimes B'$, and where $L_1^* = J_1'\Gamma_B(L) + \Gamma_B^+(L_1)\otimes B'$.

In the ensuing calculation, we shall drop the subscript "1", and write L for L_1, etc. Furthermore, if $x \in L$, $x^{[n]}$ will denote $\varphi(x)^{[n]}$, where $\varphi: L \to \Gamma_B^+(L)$ is the canonical map. Recall from (3.19) that $K = (K_1 + K_2)$, where K_1 is the ideal generated by $\{x^{[1]} - x\colon x \in L\}$ and K_2 is the ideal generated by $\{y^{[n]} - \bar{\delta}_n(y)^{[1]}\colon y \in J\}$. Recall also that $K^+ = K \cap \Gamma_B^+(L)$ is a *sub-PD ideal of* $\Gamma_B^+(L)$ *(this was the key to the D-construction)*. It follows that $K'^+ = K'\cap(\Gamma_B^+(L)\otimes B')$ is a sub-PD ideal of $\Gamma_B^+(L)\otimes B'$, since K'^+ equals the image of $K^+\otimes B'$.

Because $K_2' \subseteq \Gamma_B^+(L)\otimes B' \subseteq L^*$, we have $K' \cap L^* = K_1' \cap L^* + K_2'$. Now if $x_2 \in K_2'$, $x_2 \in K'^+ \cap L^*$, so $x_2^{[n]} \subseteq K' \cap L^*$ if $n \geq 1$. If we can prove the same for $x_1 \in K_1' \cap L^*$, we are finished, using the formula for $(x_1 + x_2)^{[n]}$.

Suppose $x \in K_1' \cap L^*$, say $x = \Sigma(x_i^{[1]} - x_i)c_i$, with $c_i \in \Gamma_B(L)\otimes B'$. If we write $c_i = c_i^{(0)} + c_i^+$, with $c_i^{(0)} \in \Gamma_B^0(L)\otimes B'$ and $c_i^+ \in \Gamma_B^+(L)\otimes B'$, we get a similar decomposition: $x = x^{(0)} + x^+$, with x^+ lying in the sub-PD ideal K'^+ of $\Gamma_B^+(L)\otimes B'$. Thus it suffices to consider the term $x^{(0)}$; i.e. we may assume that $x = \Sigma(x_i^{[1]} - x_i)\otimes b_i'$, with $b_i' \in B'$.

Since $B'/J' \cong B/J$, B' is generated as a B-module by 1 and J' . It follows that x can be rewritten as a sum $(x_0^{[1]} - x_0)\otimes 1 + \Sigma(x_i^{[1]} - x_i)\otimes\lambda_i$, where $\lambda_i \in J'$. Notice that each term $(x_i^{[1]} - x_i)\otimes\lambda_i$ belongs to $L^* = J' + \Gamma^+ \otimes B'$, and since x also belongs to L^*, so does $(x_0^{[1]} - x_0)\otimes 1$. Since the terms $(x_i^{[1]} - x_i)\otimes\lambda_i$ belong to the sub PD ideal $K'L^*$ of L^*, they cause no difficulty.

It remains to check that if $(x_0^{[1]} - x_0)\otimes 1$ lies in $K' \cap L^*$, so do its divided powers. Since $L^* = J' + \Gamma_B^+(L)\otimes B'$ and $(x_0^{[1]} - x_0)\otimes 1 \in L^*$, we see that in fact $x_0\otimes 1 \in J'$, and since $B/J \to B'/J'$ is an isomorphism, it follows that $x_0 \in J$. This, together with the fact that $(B,J,\delta) \to (B',J',\delta')$ is a PD morphism, allows us to make the following computation (in which we drop the useless subscript).

$$((x^{[1]} - x)\otimes 1)^{[n]} = \sum_{r+s=n} (x^{[1]}\otimes 1)^{[r]}(-1)^s \delta_s'(x\otimes 1)$$

$$= \sum_{r+s=n} x^{[r]}(-1)^s \delta_s(x)\otimes 1$$

$$= \sum_{r+s=n} x^{[r]}(-1)^s (\delta_s(x) - x^{[s]})\otimes 1 +$$

$$+ \sum_{r+s=n} x^{[r]}(-1)^s x^{[s]}\otimes 1 \quad .$$

In the last line, the terms $\delta_s(x)-x^{[s]}$ belong to the ideal K, and the second term is zero. It follows that $((x^{[1]} - x)\otimes 1)^{[n]} \in K' \cap L^*$, as desired. □

6.3 <u>Corollary.</u> The canonical maps:

$$\mathcal{D}_Y(X)\otimes_{\mathcal{O}_X}\mathcal{D}_{X/S}(1) \to \mathcal{D}_Y(X\underset{S}{\times}X) \text{ and } \mathcal{D}_{X/S}(1)\otimes_{\mathcal{O}_X}\mathcal{D}_Y(X) \to \mathcal{D}_Y(X\underset{S}{\times}X)$$

are isomorphisms.

<u>Proof.</u> We have diagrams:

$$\begin{array}{ccc} X & \longrightarrow & X \underset{S}{\times} X \\ i\uparrow & & p_1\downarrow\ \downarrow p_2 \\ Y & \longrightarrow & X \end{array} \qquad\qquad \begin{array}{ccc} & \nearrow & D_{X/S}(1) \\ & & \pi_1\downarrow\ \downarrow\pi_2 \\ X & \longrightarrow & X \end{array}$$

The morphisms π_i are morphisms in Cris(X/S), and the fact that $i_{cris_*}(\mathcal{O}_{Y/S})$ is a crystal tells us that the maps:

$$\pi_i^* \, i_{cris_*}(\mathcal{O}_{Y/S})_X \to i_{cris_*}(\mathcal{O}_{Y/S})_{\mathcal{D}_{X/S}(1)}$$

are isomorphisms. Recalling that

$i_{cris_*}(\mathcal{O}_{Y/S})_X \cong \mathcal{D}_{Y,\gamma}(X)$ and $i_{cris_*}(\mathcal{O}_{Y/S})_{\mathcal{D}_{X/S}}(1) \cong \mathcal{D}_{Y,\overline{[\ ]}}(\mathcal{D}_{X/S}(1))$, we are reduced to seeing that $\mathcal{D}_{Y,\overline{[\ ]}}(\mathcal{D}_{X/S}(1)) \cong \mathcal{D}_{Y,\gamma}(X\underset{S}{\times}X)$. This follows easily from the universal mapping properties. □

6.4 <u>Exercise.</u> Deduce from the Corollary a natural isomorphism $\varepsilon: \mathcal{D}_{X/S}(1)\otimes_{\mathcal{O}_X}\mathcal{D}_Y(X) \to \mathcal{D}_Y(X)\otimes_{\mathcal{O}_X}\mathcal{D}_{X/S}(1)$. Show that ε is an HPD stratification on $\mathcal{D}_Y(X)$. If $\nabla: \mathcal{D}_Y(X) \to \mathcal{D}_Y(X)\otimes\Omega^1_{X/S}$ is the corresponding integrable connection, show that $\nabla(y^{[k]}) = y^{[k-1]}\otimes dy$, for any section y of the ideal of Y in X. The complex $\mathcal{D}_{Y,\gamma}(X)\otimes\Omega^{\cdot}_{X/S}$ can be thought of as the deRham complex of $\mathcal{D}_{Y,\gamma}(X)$, and we shall sometimes denote it by $\Omega^{\cdot}_{\mathcal{D}_{Y,\gamma}(X)/S}$.

6.5 <u>Exercise.</u> Suppose $f: X' \to X$ is a morphism covering the PD morphism $(S',I',\gamma) \to (S,I,\gamma)$, and E is a crystal on X/S. Then $f^*_{cris}(E)$ is a crystal on X'/S'. If $h: T' \to T$ is a morphism from an object in Cris(X'/S') to an object in Cris(X/S), $f^*_{cris}(E)_{T'} \cong h^*(E_T)$.

Let us now change our point of view and notation. We want to study crystals over an S-scheme X which is not smooth by embedding X in some smooth S-scheme Y.

6.6 <u>Theorem.</u> If $X \hookrightarrow Y$ is a closed immersion of S-schemes, with Y/S smooth, the following categories are naturally equivalent:

(i) The category of crystals of $\mathcal{O}_{X/S}$-modules on Cris(X/S).

(ii) The category of $\mathcal{D}_{X,\gamma}(Y)$-modules with an HPD stratification (as an $\mathcal{O}_Y$-module) which is compatible with the canonical HPD stratification (6.4) of $\mathcal{D}_{X,\gamma}(Y)$.

(iii) The category of $\mathcal{D}_{X,\gamma}(Y)$-modules with an integrable, quasi-nilpotent connection (as an $\mathcal{O}_Y$-module) which is compatible with the canonical connection on $\mathcal{D}_{X,\gamma}(Y)$.

Before giving the proof, let me remark that in (iii), compatibility of a connection ∇_E on a $\mathcal{D}_{X,\gamma}(Y)$-module E with the connection $\nabla_{\mathcal{D}}$ on $\mathcal{D}_{X,\gamma}(Y)$ means that if α is a section of $\mathcal{D}$ and e a section of E, $\nabla_E(\alpha e) = \nabla_{\mathcal{D}}(\alpha)\otimes e + \alpha\nabla_E(e)$, under the canonical identification $E\otimes_{\mathcal{O}_Y}\Omega^1_{Y/S} \cong E\otimes_{\mathcal{D}}\mathcal{D}\otimes\Omega^1_{Y/S}$. Thus if we (abusively) view $\mathcal{D}\otimes\Omega^1_{Y/S}$ as $\Omega^1_{\mathcal{D}/S}$ and $\nabla_{\mathcal{D}}$ as the exterior derivative, we can view ∇_E as a connection on the $\mathcal{O}_{\mathcal{D}}$-module E in the usual sense.

Proof of 6.6. It suffices to check the equivalence of (i) and (ii), and to refer to (4.12) for the equivalence of (ii) and (iii).

Suppose E is a crystal on Cris(X/S). There are two morphisms in Cris(X/S) $p_i\colon \mathcal{D}_{X,\gamma}(Y\underset{S}{\times}Y) \to \mathcal{D}_{X,\gamma}(Y)$, and since E is a crystal, we get isomorphisms $p_i^*(E_{\mathcal{D}_{X,\gamma}(Y)}) \to E_{\mathcal{D}_{X,\gamma}(Y\underset{S}{\times}Y)}$. Combining these, we get an isomorphism:

$\mathcal{D}_{X,\gamma}(Y\underset{S}{\times}Y)\otimes E_{D_{X,\gamma}(Y)} \to E_{D_{X,\gamma}(Y)}\otimes\mathcal{D}_{X,\gamma}(Y\underset{S}{\times}Y)$. (The tensor product is taken over $\mathcal{D}_{X,\gamma}(Y)$.) Since there are isomorphisms (6.4): $\mathcal{D}_{X,\gamma}(Y\underset{S}{\times}Y) \cong \mathcal{D}_X(Y)\otimes_{\mathcal{O}_Y}\mathcal{D}_{Y/S}(1) \cong \mathcal{D}_{Y/S}(1)\otimes_{\mathcal{O}_Y}\mathcal{D}_X(Y)$, we can interpret the above as an isomorphism:

$\mathcal{D}_{Y/S}(1)\otimes_{\mathcal{O}_Y} E_{\mathcal{D}_{X,\gamma}(Y)} \xrightarrow{\varepsilon} E_{\mathcal{D}_{X,\gamma}(Y)}\otimes_{\mathcal{O}_Y}\mathcal{D}_{Y/S}(1)$, in other words, as an HPD stratification on $F = E_{D_{X,\gamma}(Y)}$. Notice that the HPD stratification on $\mathcal{D}_{Y/S}(1)$ is built into the construction of ε ; this insures the compatibility we claimed.

Conversely, given a $\mathcal{D}_{X/S}(Y)$-module F with compatible $\mathcal{O}_Y$-module stratification, we construct a crystal E on Cris(X/S) as follows: It suffices to specify E_T for sufficiently small $(U,T,\delta) \in$ Cris(X/S), e.g. if there exists an S-morphism $h: T \to D_{X,\gamma}(Y)$ extending $U \to Y$ (c.f. (5.28)). Viewing F as a sheaf on $D_{X,\gamma}(Y)$, we define E_T to be $h^*(F)$. The fact that E_T does not depend on h, up to canonical isomorphism, comes from the HPD stratification on F, viewed as an isomorphism $\mathcal{D}_{X,\gamma}(Y\underset{S}{\times}Y)\otimes_{\mathcal{D}_{X,\gamma}(Y)}F \leftrightarrow F\otimes_{\mathcal{D}_{X,\gamma}(Y)}\mathcal{D}_{X,\gamma}(Y\underset{S}{\times}Y)$. This part of the argument is the same as it was without divided powers (2.10ff), so we do not repeat it. □

6.7 <u>Corollary.</u> Suppose we have a Cartesian square, where X/S is locally of finite type and $(S_0,I_0,\gamma_0)\hookrightarrow(S,I,\gamma)$ is a PD immersion defined by a sub-PD ideal of I . Then the natural map ((Crystals on X/S)) → ((Crystals on X_0/S)) is an equivalence of categories.

$$\begin{array}{ccc} X_0 & \xhookrightarrow{i} & X \\ \downarrow & & \downarrow \\ S_0 & \hookrightarrow & S \end{array}$$

<u>Proof.</u> Since sheaves can be determined and even constructed locally, the question is of a local nature on X. Hence we may

assume that X can be embedded in a scheme Y which is smooth over S and to which γ also extends. In (5.17) we showed that $i_{cris_*}(\mathcal{O}_{X_0/S}) \cong \mathcal{O}_{X/S}$, and in (6.2) we showed that $j_{cris_*}(\mathcal{O}_{X/S})_Y \cong \mathcal{D}_{X,\gamma}(Y)$, where $j\colon X \to Y$ is our embedding. Applying (6.2) with $j_0\colon X_0 \to Y = j \circ i$, we see that $\mathcal{D}_{X_0,\gamma_0}(Y) \cong \mathcal{D}_{X,\gamma}(Y)$. Now using description (6.6) (ii) or (iii) of crystals, we get the corollary immediately. This proof also proves the next corollary. □

6.8 Corollary. Assume also in the situation of (6.7) that X/S is smooth. Then the category of crystals on Cris(X_0/S) is equivalent to the category of $\mathcal{O}_X$-modules with integrable, quasi-nilpotent connection relative to S. □

We shall now indicate the divided power analogue of the linearization of differential operators we sketched in (2.14), a construction we shall use in the next section to calculate the cohomology of a crystal. If Y is an S-scheme (to which γ , as always, extends), and if E is a sheaf of $\mathcal{O}_Y$-modules, we define $L_Y(E)$ to be $\mathcal{D}_{Y/S}(1)\otimes_{\mathcal{O}_Y} E$, with the $\mathcal{O}_Y$-module structure from the left. An HPD differential operator $u\colon E \longrightarrow F$ will induce an $\mathcal{O}_Y$-linear map $L_Y(u)\colon L_Y(E) \longrightarrow L_Y(F)$, as follows: u is an $\mathcal{O}_Y$-linear map: $\mathcal{D}_{Y/S}(1)\otimes_{\mathcal{O}_Y} E \longrightarrow F$, and $L_Y(u)$ is defined to be the composition:

$$L_Y(u)\colon \mathcal{D}_{Y/S}(1)\otimes E \xrightarrow{\delta\otimes id_E} \mathcal{D}_{Y/S}(1)\otimes\mathcal{D}_{Y/S}(1)\otimes E \xrightarrow{id_{\mathcal{D}}\otimes u} \mathcal{D}_{Y/S}(1)\otimes F \quad .$$

This makes L_Y into a functor.

Just as before $L_Y(E)$ has a canonical HPD stratification: $\mathcal{D}_{Y/S}(1)\otimes_{\mathcal{O}_Y} L_Y(E) \to L_Y(E)\otimes_{\mathcal{O}_Y}\mathcal{D}_{Y/S}(1)$, induced from the map given in (2.14) Here is another description: The left and right $\mathcal{O}_Y$-module structures on $\mathcal{D}_{Y/S}(1)$ $\mathcal{O}_Y$-modules $p_{1*}\mathcal{D}_{Y/S}(1)$ and $p_{2*}\mathcal{D}_{Y/S}(1)$, respectively), each have a canonical HPD stratification, the second of which corresponds to δ (c.f. (2.14)). If E is an $\mathcal{O}_Y$-module, $L_Y(E) = \mathcal{D}_{Y/S}(1)\otimes E$ is computed using the right structure, and it inherits an HPD stratification (and an $\mathcal{O}_Y$-module structure) from the left-structure. If $u\colon E \to F$ is an HPD differential operator, $L_Y(u)\colon L_Y(E) \to L_Y(F)$ is the map induced from the stratification on $p_{2*}\mathcal{D}_{Y/S}(1)$ via (4.8) (it is an HPD differential operator using the right structure of $L_Y(E)$, but is $\mathcal{O}_Y$-linear and even <u>horizontal</u> using the left structure. In summary:

6.9 <u>Construction.</u> There is a functor L_Y from the category of $\mathcal{O}_Y$-modules and HPD differential operators to the category of HPD stratified $\mathcal{O}_Y$-modules and $\mathcal{O}_Y$-linear horizontal maps. If Y/S is smooth, $\mathcal{D}_{Y/S}(1)$ is locally free by the remark in (4.1), and hence L_Y is exact. □

The L_Y construction furnishes us with a large collection of HPD stratifications, and hence with a large collection of crystals, via (6.6). More generally, if $i\colon X \to Y$ is a closed immersion and Y/S is smooth, we shall use the following notation: If E is a sheaf of $\mathcal{O}_Y$-modules, $L_Y(E)$ will be the sheaf of $\mathcal{O}_Y$-modules with HPD stratification indicated above. $\mathrm{L}_Y(E)$ will be the crystal we get on Y/S by construction (6.6), and $\mathrm{L}_i(E)$ (or just $\mathrm{L}(E)$) will be $i^*_{cris}\mathrm{L}_Y(E)$, a crystal on X/S by (6.5).

We can give another description of $L(E)$ in terms of our localization diagram (5.24). Let $\tilde{Y} = i^*(Y)$ in $(X/S)_{cris}$ (represented, the reader will recall, by $D_{X,\gamma}(Y)$). We have a diagram:

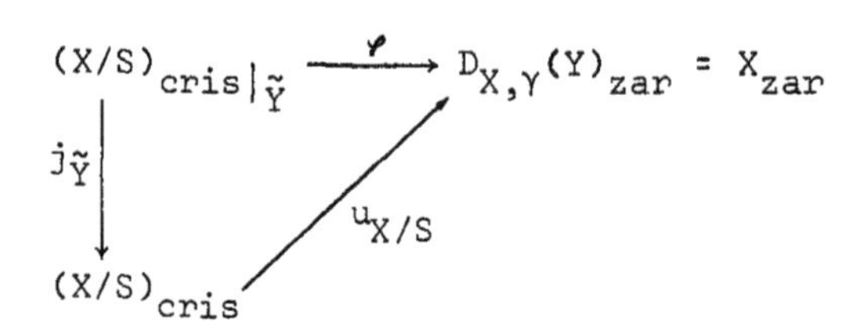

$$\begin{array}{ccc} (X/S)_{cris}|_{\tilde{Y}} & \xrightarrow{\varphi} & D_{X,\gamma}(Y)_{zar} = X_{zar} \\ \downarrow j_{\tilde{Y}} & \nearrow u_{X/S} & \\ (X/S)_{cris} & & \end{array}$$

6.10 <u>Proposition.</u> If E is a sheaf of $\mathcal{O}_Y$-modules and $\lambda\colon D_{X,\gamma}(Y)_{zar} \to Y_{zar}$ is the natural map, there is a natural isomorphism:

$$L(E) \longrightarrow j_{\tilde{Y}_*}(\lambda\circ\varphi)^*E$$

<u>Proof.</u> An arrow $L(E) \to j_{\tilde{Y}_*}(\lambda\circ\varphi)^*E$ is equivalent to an arrow $j_{\tilde{Y}}{}^*L(E) \to (\lambda\circ\varphi)^*E \cong \varphi^*(\lambda^*(E))$. Using (5.25.1) we see that the source of this hoped for arrow assigns to any $u\colon T \to D_{X,\gamma}(Y)$ the sheaf $L(E)_T$, and by (5.26) its target assigns to u the sheaf $u^*\lambda^*(E)$. Now $L(E)_T$ is by definition $i_{cris}{}^*(L_Y(E))_T$, and by (6.5) we see that this is $(\lambda\circ u)^*(L_Y(E))$. Thus we have only to give a compatible collection of maps $u^*\lambda^*L_Y(E) \to u^*\lambda^*(E)$, for every u, and a map $L_Y(E) \to E$ will certainly do. But $L_Y(E) = D_{Y/S}(1)\otimes E$, and we can just take the map induced by $a\otimes b\otimes x \longmapsto abx$.

To show that our arrow $L(E) \to j_{\tilde{Y}_*}(\lambda\circ\varphi)^*E$ is an isomorphism, recall from (5.25.2) that if $T \in \mathrm{Cris}(X/S)$, $(j_{\tilde{Y}_*}(\lambda\circ\varphi)^*E)_T = p_{T_*}((\lambda\circ\varphi)^*E)_{P_D}$, where $p_T\colon D_{X,\gamma}(T\times_S Y) \to T$ and

$p_D\colon D_{X,\gamma}(T\underset{S}{\times}Y) \to \tilde{Y} = D_{X,\gamma}(Y)$ are the natural maps. Thus, we have

$(j_{\tilde{Y}_*}(\lambda\circ\varphi)^*E)_T \cong p_{T*}p_Y{}^*(E)$, where $p_Y\colon D_{X,\gamma}(T\underset{S}{\times}Y) \to Y$ is the natural projection. On the other hand, if there is a map $T \to \tilde{Y}$ (which we may assume, by (5.28)), $L(E)_T$ is just $\mathcal{O}_T\otimes_{\mathcal{O}_Y}\mathcal{D}_{Y/S}(1)\otimes E$. It is clear that we are thus reduced to proving that the map $\mathcal{O}_T\otimes\mathcal{D}_{Y/S}(1) \to \mathcal{D}_{X,\gamma}(T\underset{S}{\times}Y)$ is an isomorphism. The problem is to show that the ideal of X in $\mathcal{O}_T\otimes\mathcal{D}_{Y/S}(1)$ has a PD structure compatible with γ. This ideal is $J\otimes\mathcal{D}_{Y/S}(1) + (\mathcal{O}_T\otimes K)$, where $J \subseteq \mathcal{O}_T$ is the ideal of $\mathcal{O}_X$ and $K \subseteq \mathcal{D}_{Y/S}(1)$ is the ideal of Y — an augmentation ideal. Since $\mathcal{D}_{Y/S}(1)$ is flat, the PD structure δ of J extends to a PD structure δ' on $J\otimes\mathcal{D}_{Y/S}(1)$, and since K is an augmentation ideal, its PD structure is compatible with δ'. This gives us the desired extension, establishes the isomorphism $\mathcal{O}_T\otimes_{\mathcal{O}_Y}\mathcal{D}_{Y/S}(1) \to \mathcal{D}_{X,\gamma}(T\underset{S}{\times}Y)$, and proves the proposition. □

6.10.1 <u>Remark.</u> Actually, we could have started with a sheaf E of $\mathcal{D}$-modules instead of a sheaf of $\mathcal{O}_Y$-modules. Then if we let $L(E) = j_{\tilde{Y}*}\varphi^*(E)$, it can be shown that $L(E)$ is a crystal. If $T \in \mathrm{Cris}(X/S)$, $L(E)_T \cong p_{T*}p_D{}^*(E)$, where p_T and p_D are the maps from $D_{X,\gamma}(T\underset{S}{\times}Y)$ to T and to $D_{X,\gamma}(Y)$, respectively.

Let us apply the functor L_Y to the complex $(\Omega^{\bullet}_{Y/S},d)$ of $\mathcal{O}_Y$-modules and HPD differential operators. To see that we get a complex $L_Y(\Omega^{\bullet}_{Y/S})$, the reader can show that the composition

$\Omega^{i-1}_{Y/S} \to \Omega^{i}_{Y/S} \to \Omega^{i+1}_{Y/S}$ is zero as an HPD differential operator.
We shall take a more explicit approach, attempting to demystify the L_Y construction by working it out in detail using local coordinates. (All we are really doing is Exercise 6.4 in a special case.)

6.11 Lemma. Suppose $x_1 \ldots x_n$ are local coordinates on Y, $\xi_i = 1\otimes x_i - x_i \otimes 1$ in $\mathcal{D}_{Y/S}(1)$, ω is a section of $\Omega^k_{Y/S}$, and a is a section of $\mathcal{O}_Y$. Then the map
$L_Y(d)\colon \mathcal{D}_{Y/S}(1)\otimes\Omega^k_{Y/S} \to \mathcal{D}_{Y/S}(1)\otimes\Omega^{k+1}_{Y/S}$ behaves as follows:

$$L_Y(d)(a\xi_1^{[k_1]}\cdots\xi_n^{[k_n]}\otimes\omega) = a\sum_{i=1}^{n}\xi_1^{[k_1]}\cdots\xi_i^{[k_i-1]}\cdots\xi_n^{[k_n]}\otimes dx_i\wedge\omega$$
$$+ a\xi_1^{[k_1]}\cdots\xi_n^{[k_n]}\otimes d\omega \quad .$$

Proof. Recall that $\delta(\xi_i) = 1\otimes\xi_i + \xi_i\otimes 1$ in $\mathcal{D}_{Y/S}(1)\otimes\mathcal{D}_{Y/S}(1)$. Since d is a differential operator of order one, we have a diagram:

$$\begin{array}{ccccc}
\mathcal{D}_{Y/S}(1)\otimes\Omega^k_{Y/S} & \xrightarrow{\delta\otimes \mathrm{id}_\Omega} & \mathcal{D}_{Y/S}(1)\otimes\mathcal{D}_{Y/S}(1)\otimes\Omega^k_{Y/S} & \longrightarrow & \mathcal{D}_{Y/S}(1)\otimes\Omega^{k+1}_{Y/S} \\
 & & \downarrow{\scriptstyle \mathrm{id}_{\mathcal{D}}\otimes\pi\otimes\mathrm{id}_\Omega} & \nearrow{\scriptstyle \mathrm{id}_{\mathcal{D}}\otimes\tilde{d}} & \\
 & & \mathcal{D}_{Y/S}(1)\otimes\mathcal{D}^1_{Y/S}\otimes\Omega^k_{Y/S} & &
\end{array}$$

In this diagram, $L_Y(d)$ is the composition of the two horizontal arrows, $\pi\colon \mathcal{D}_{Y/S}(1) \to \mathcal{D}^1_{Y/S}$ is the natural projection, and $\tilde{d}\colon \mathcal{D}^1_{Y/S}\otimes\Omega^k_{Y/S} \to \Omega^{k+1}_{Y/S}$ is the $\mathcal{O}_Y$-linearization of d. Now since

δ is a PD morphism, we compute:

$$\delta\otimes id_\Omega(a\ \xi^{[k]}\otimes\omega) = a\ \delta(\xi)^{[k]}\otimes\omega$$
$$= a(1\otimes\xi + \xi\otimes 1)^{[k]}\otimes\omega$$
$$= a\cdot\sum_{i+j=k}\xi^{[i]}\otimes\xi^{[j]}\otimes\omega\ .$$

Recalling that π kills $\xi^{(\ell)}$ if $|\ell|\geq 2$, we see that $id_{\mathcal{D}}\otimes\pi\otimes id_\Omega$ maps this to $a\cdot\sum_{i=1}^{n}\xi^{[k-1_i]}\otimes\xi_i\otimes\omega + a\xi^{[k]}\otimes\omega$. Since $\tilde{d}(1\otimes\omega) = d\omega$ and $\tilde{d}(\xi_i\otimes\omega) = dx_i\wedge\omega$, the formula follows. □

It is not hard to deduce from the above lemma the fact that $L_Y(\Omega^{\boldsymbol{\cdot}}_{Y/S})$ is indeed a complex. In fact, it will enable us to prove the Poincaré Lemma in crystalline cohomology.

6.12 Theorem. Suppose Y/S is smooth and $i\colon X\hookrightarrow Y$ is a closed immersion. Then the complex $L(\Omega^{\boldsymbol{\cdot}}_{Y/S}) \underset{def}{=} i^*_{cris}\ L_Y(\Omega^{\boldsymbol{\cdot}}_{Y/S}))$ of crystals on X/S is a resolution of $\mathcal{O}_{X/S}$. That is, there is a natural quasi-isomorphism: $\mathcal{O}_{X/S}\to L(\Omega^{\boldsymbol{\cdot}}_{Y/S})$ of complexes of abelian sheaves in $(X/S)_{cris}$.

Proof. There is a natural map: $\mathcal{O}_Y\to\mathcal{D}_{Y/S}(1)\cong L_Y(\mathcal{O}_Y)$ sending x to $x\otimes 1$, which is compatible with HPD stratifications. (The reader may easily check this from the fact that the HPD stratification on $\mathcal{O}_Y$ corresponds to the identity map $\mathcal{D}_{Y/S}(1)\to\mathcal{D}_{Y/S}(1)$, and that the one on $\mathcal{D}_{Y/S}(1)$ corresponds to the map $\mathcal{D}_{Y/S}(1)\otimes\mathcal{D}_{Y/S}(1)\to\mathcal{D}_{Y/S}(1)\otimes\mathcal{D}_{Y/S}(1)$ coming from $a\otimes b\otimes c\otimes d\longrightarrow 1\otimes d\otimes a\otimes bc$.) It follows immediately from the local coordinate calculation (6.11) that the composition $\mathcal{O}_Y \to L_Y(\mathcal{O}_Y) \to L_Y(\Omega^1_{Y/S})$ is zero, and hence that we have a morphism of complexes $\mathcal{O}_Y \to L^{\cdot}_Y(\Omega^{\cdot}_{Y/S})$.

We get induced a map of complexes of crystals on Y/S: $\mathcal{O}_{Y/S} \to L^{\cdot}_Y(\Omega^{\cdot}_{Y/S})$ and hence by applying i^*_{cris}, a map of complexes of crystals on X/S: $\mathcal{O}_{X/S} \to L^{\cdot}(\Omega^{\cdot}_{Y/S})$. To show that $\mathcal{O}_{X/S} \to L^{\cdot}(\Omega^{\cdot}_{Y/S})$ is a quasi-isomorphism, we may compute on a small $(U,T,\delta) \in \mathrm{Cris}(X/S)$ which admits a section $h: T \to Y$. By exercise (6.5), $L(\Omega^{\cdot}_{Y/S})_T \cong h^*(L(\Omega^{\cdot}_{Y/S})_Y) \cong h^*(L_Y(\Omega^{\cdot}_{Y/S}))$. Assume further that $h(T)$ is contained in an open set of Y on which $x_1 \dots x_n$ are local coordinates. Then on this open set, $\mathcal{D}_{X/S}(1)$ is isomorphic to the PD polynomial algebra $\mathcal{O}_Y\langle\xi_1 \dots \xi_n\rangle$, by (3.32). Thus, $L(\Omega^{\cdot}_{Y/S})_T$ is just the deRham complex of the PD polynomial algebra $\mathcal{O}_T\langle\xi_1 \dots \xi_n\rangle$, relative to T. (This can be made explicit from the formula (6.11).) Hence we are reduced to proving that this complex is a resolution of $\mathcal{O}_T$. If $n = 1$ this is clear; the complex is just $\mathcal{O}_T \to \mathcal{O}_T\langle\xi\rangle \xrightarrow{d} \mathcal{O}_T\langle\xi\rangle dx$ with $d(\xi^{[i]}) = \xi^{[i-1]}dx$. It is simplest to proceed by induction on n. Assuming that $\mathcal{O}_T \to \Omega^{\cdot}_{\mathcal{O}_T\langle\xi_1 \dots \xi_{n-1}\rangle}$ is a quasi-isomorphism, so is $\mathcal{O}_T \otimes \Omega^{\cdot}_{\mathcal{O}_T\langle\xi_n\rangle} \to \Omega^{\cdot}_{\mathcal{O}_T\langle\xi_1 \dots \xi_{n-1}\rangle} \otimes_{\mathcal{O}_T} \Omega^{\cdot}_{\mathcal{O}_T\langle\xi_n\rangle}$, because $\Omega^{\cdot}_{\mathcal{O}_T\langle\xi_n\rangle}$ consists of locally free $\mathcal{O}_T$-modules. But this arrow can be identified with the arrow: $\Omega^{\cdot}_{\mathcal{O}_T\langle\xi_n\rangle} \to \Omega^{\cdot}_{\mathcal{O}_T\langle\xi_1 \dots \xi_n\rangle}$, and since the arrow $\mathcal{O}_T \to \Omega^{\cdot}_{\mathcal{O}_T\langle\xi_n\rangle}$ is a quasi-isomorphism, we are done. □

A refinement of the above result will provide us with a beautiful crystalline interpretation of the Hodge filtration.

Let us keep the notation of the previous theorem.

We begin by restricting the exact sequence (5.2.3)

$0 \to J_{X/S} \to \mathcal{O}_{X/S} \to i_{X/S*}\mathcal{O}_X \to 0$ to $(X/S)_{cris|\tilde{Y}}$, obtaining:

$$0 \to J_{X/S|\tilde{Y}} \to \mathcal{O}_{X/S|\tilde{Y}} \to j^*_{\tilde{Y}}\, i_{X/S*}\mathcal{O}_X \to 0 \quad .$$

Since $j_{\tilde{Y}*}$ is exact (5.27.1), since $j_{\tilde{Y}*}\, j^*_{\tilde{Y}}\, i_{X/S*}\mathcal{O}_X$ is just $i_{X/S*}\mathcal{O}_X$ again, and since $j_{\tilde{Y}*}(\mathcal{O}_{X/S|\tilde{Y}})$ is $L(\mathcal{O}_Y)$ (6.10), we obtain an exact sequence:

$$(6.13.1) \qquad 0 \to j_{\tilde{Y}*}(J_{X/S|\tilde{Y}}) \to L(\mathcal{O}_Y) \to i_{X/S*}\mathcal{O}_X \to 0 \quad .$$

It is clear that $K = j_{\tilde{Y}*}(J_{X/S|\tilde{Y}})$ is a P.D. ideal in $L(\mathcal{O}_Y)$. In terms of local coordinates, if $L(\mathcal{O}_Y)_T = \mathcal{O}_T\langle\xi_1 \ldots \xi_n\rangle$, K is the ideal generated by $\xi_1 \ldots \xi_n$ and the ideal $(J_{X/S})_T$ of $X \cap T$ in T. Let $K^{[\cdot]}$ denote the PD filtration (3.24); it follows from the local formula (6.11) that $L(d)$ maps $K^{[\cdot]}L(\Omega^q_{Y/S})$ into $K^{[\cdot -1]}L(\Omega^{q+1}_{X/S})$. Set $F^m_X L(\Omega^q_{Y/S}) = K^{[m-q]}\, L(\Omega^q_{Y/S})$, so that $F^m_X L(\Omega^{\cdot}_{Y/S})$ is a subcomplex of $L(\Omega^{\cdot}_{Y/S})$. Noting that the canonical map $\mathcal{O}_{X/S} \to L(\mathcal{O}_Y)$ sends $J_{X/S}$ into K, we see that we have an augmentation: $J^{[m]}_{X/S} \longrightarrow F^m_X L(\Omega^{\cdot}_{Y/S})$.

6.13 <u>Theorem.</u> The above map is a quasi-isomorphism:

$$J^{[m]}_{X/S} \longrightarrow F^m_X L(\Omega^{\cdot}_{Y/S}) \quad .$$

<u>Proof.</u> First let us describe the ideal $K^{[m]}_T$ in terms of local coordinates, where $T = (U,T,\delta)$. I claim that it consists of the set of PD polynomials $\Sigma a_k \xi^{[k]}$ such that $a_k \in J_T^{[m-|k|]}$

for all k. It is easy to see that this set is a sub PD ideal of $K_T^{[m]}$. Hence it suffices to check that it contains a set of PD generators of $K_T^{[m]}$, for instance $\{\delta_{i_1}(\alpha_1)\cdots\delta_{i_r}(\alpha_r) : \Sigma i_j \geq m,$ $\alpha_i \in J_T \cup \{\xi_1 \ldots \xi_n\}$. This is clear.

In particular, if $n = 1$, $K_T^{[m]}$ is the set of PD polynomials $\Sigma a_k \xi^{[k]}$ such that $a_k \in J_T^{[m-k]}$ for all k. The inclusion of complexes: $i: \mathcal{O}_T \to \Omega^{\bullet}_{\mathcal{O}_T<\xi>}$ sends $J_T^{[m]}$ to $F^m\Omega^{\bullet}_{\mathcal{O}_T<\xi>}$. There is a section $\pi: \Omega^{\bullet}_{\mathcal{O}_T<\xi>} \to \mathcal{O}_T$, (the $\mathcal{O}_T$-linear mapping sending $\xi^{[k]}$ to zero for $k \geq 1$) which sends $F^m\Omega^{\bullet}_{\mathcal{O}_T<\xi>}$ to $J_T^{[m]}$, and a chain homotopy $s: i\circ\pi \sim id_{\Omega}$, (the $\mathcal{O}_T$-linear map sending $\xi^{[k]}d\xi$ to $\xi^{[k+1]}$) under which $F^m\Omega^{\bullet}_{\mathcal{O}_T<\xi>}$ is invariant. Thus, when $n = 1$, the result is clear.

To proceed by induction on n we have to show that the map $i_n : F^m\Omega^{\bullet}_{\mathcal{O}_T<\xi_1 \ldots \xi_{n-1}>} \to F^m\Omega^{\bullet}_{\mathcal{O}_T<\xi_1 \ldots \xi_n>}$ is a quasi-isomorphism. Recall that $\Omega^{\bullet}_{\mathcal{O}_T<\xi_1 \ldots \xi_n>} \cong \Omega^{\bullet}_{\mathcal{O}_T<\xi_1 \ldots \xi_{n-1}>} \otimes \Omega^{\bullet}_{\mathcal{O}_T<\xi>}$; then i_n can be identified with:

$$id_{\Omega^{\bullet}_{\mathcal{O}_T<\xi_1 \ldots \xi_{n-1}>}} \otimes i : \Omega^{\bullet}_{\mathcal{O}_T<\xi_1 \ldots \xi_{n-1}>} \otimes \mathcal{O}_T \to \Omega^{\bullet}_{\mathcal{O}_T<\xi_1 \ldots \xi_{n-1}>} \otimes \Omega^{\bullet}_{\mathcal{O}_T<\xi>} .$$

All that remains is to observe that the projection $id\otimes\pi$ and the homotopy $id\otimes s$ are compatible with the filtration F^m. Since this is immediate, the proof follows. □

Using the same technique we can also obtain resolutions of other crystals of $\mathcal{O}_{X/S}$-modules.

6.14 Theorem. With the notations of (6.13):

6.14.1 If M is any sheaf of $\mathcal{O}_{X/S}$-modules on Cris(X/S), let $F_X^m(M \otimes_{\mathcal{O}_{X/S}} L(\Omega^q_{Y/S})) = K^{[m-q]}(M \otimes L(\Omega^q_{Y/S}))$. Then there is a canonical quasi-isomorphism:

$$J_{X/S}^{[m]}M \to F_X^m(M \otimes L(\Omega^{\cdot}_{Y/S})).$$

6.14.2 If E is a crystal of $\mathcal{O}_{X/S}$-modules and if (E,∇) is the corresponding sheaf of $\mathcal{D}_{X,Y}(Y)$-modules with integrable connection (6.6), then there is a natural quasi-isomorphism:

$$J_{X/S}^{[m]}E \to F_X^m L(E \otimes_{\mathcal{D}} \Omega^{\cdot}_{D/S}) \quad .$$

Proof. We shall leave (6.14.1) to the reader, who needs only transcribe the proof of (6.13), replacing $\mathcal{O}_X$ everywhere by M. To deduce (6.14.2), we need to find a nice quasi-isomorphism:

$$F_X^m(E \otimes L(\Omega^{\cdot}_{Y/S})) \to F_X^m L(E \otimes \Omega^{\cdot}_{Y/S}) \quad .$$

In fact, there is even an isomorphism of complexes of $L(\mathcal{O}_Y)$-modules: $L(E \otimes \Omega^{\cdot}_{Y/S}) \to E \otimes L(\Omega^{\cdot}_{Y/S})$, which, since it is linear, preserves the ideals $K^{[i]}$, hence the filtration $F_X^{\cdot}$. To find this isomorphism, we work with the associated HPD-stratified $\mathcal{O}_Y$-modules, in the following proposition.

6.15 Proposition. Let E be an $\mathcal{O}_Y$-module with HPD-stratification, and let Ω be any $\mathcal{O}_Y$-module. Then there is a canonical isomorphism: $\beta: L(E\otimes\Omega) \to E\otimes L(\Omega)$. Moreover:

6.15.1 β is compatible with the HPD-stratifications on $L(E\otimes\Omega)$ and $E\otimes L(\Omega)$.

6.15.2 If $u: \mathcal{D}(1)\otimes\Omega \to \Omega'$ is an HPD differential operator and if $\rho_E(u)$ is the induced operator $\mathcal{D}(1)\otimes E\otimes\Omega \to E\otimes\Omega'$ (4.8), then β takes $L(\rho_E(u))$ into $id_E\otimes L(u)$.

Proof. Recall that $L(E) = \mathcal{D}_{Y/S}(1)\otimes_{\mathcal{O}_Y} E$, so that β is supposed to be an isomorphism: $\mathcal{D}(1)\otimes E\otimes\Omega \to E\otimes\mathcal{D}(1)\otimes\Omega$. Evidently we can use $\beta = \varepsilon_E\otimes id_\Omega$, where ε_E is the given HPD stratification on E.

To prove that β is horizontal (i.e., that it preserves the HPD stratifications) we have to chase a tedious diagram. The essential point of the calculation is the cocycle condition for ε_E.

First we must give a more explicit formula for the HPD-stratification of the L-construction. Recall that if F is any $\mathcal{O}_Y$-module, $L(F) = \mathcal{D}_{Y/S}(1)\otimes_{\mathcal{O}_Y} F$, and $\varepsilon_L: \mathcal{D}(1)\otimes L(F) \to L(F)\otimes\mathcal{D}(1)$ is deduced from the map:

$$(\mathcal{O}_Y\otimes_{\mathcal{O}_S}\mathcal{O}_Y)\otimes_{\mathcal{O}_Y}(\mathcal{O}_Y\otimes_{\mathcal{O}_S}\mathcal{O}_Y)\otimes_{\mathcal{O}_Y} F \to (\mathcal{O}_Y\otimes_{\mathcal{O}_S}\mathcal{O}_Y)\otimes_{\mathcal{O}_Y} F\otimes_{\mathcal{O}_Y}(\mathcal{O}_Y\otimes_{\mathcal{O}_S}\mathcal{O}_Y)$$

$$(a\otimes b)\otimes(c\otimes d)\otimes x \to (1\otimes d)\otimes x\otimes a\otimes bc$$

(c.f. page 2.17).

To write this, we need the map:

$$s: \quad \mathcal{D}(1)\otimes\mathcal{D}(1) \to \mathcal{D}(1)_{\ell}: \quad (a\otimes b) \otimes (c\otimes c') \to (ac'\otimes bc)$$

Notice that this map is $\mathcal{O}_Y$-linear for the left $\mathcal{O}_Y$-module structures; to enforce this we shall sometimes use a subscript ℓ .

The reader can now easily see that ε_L is given by the following diagram:

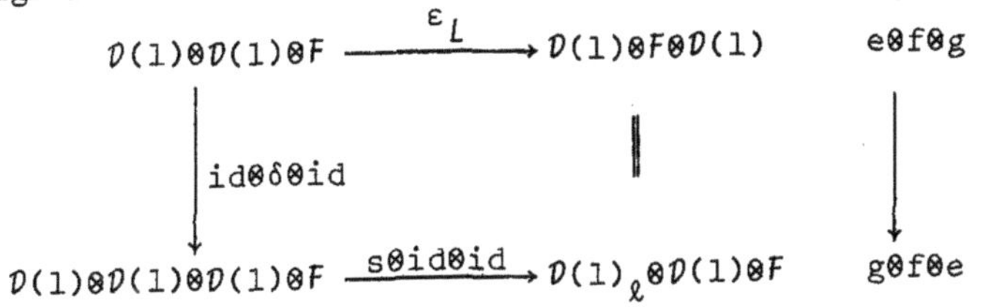

To see that the isomorphism β is compatible with the HPD-stratifications, consider the diagram below:

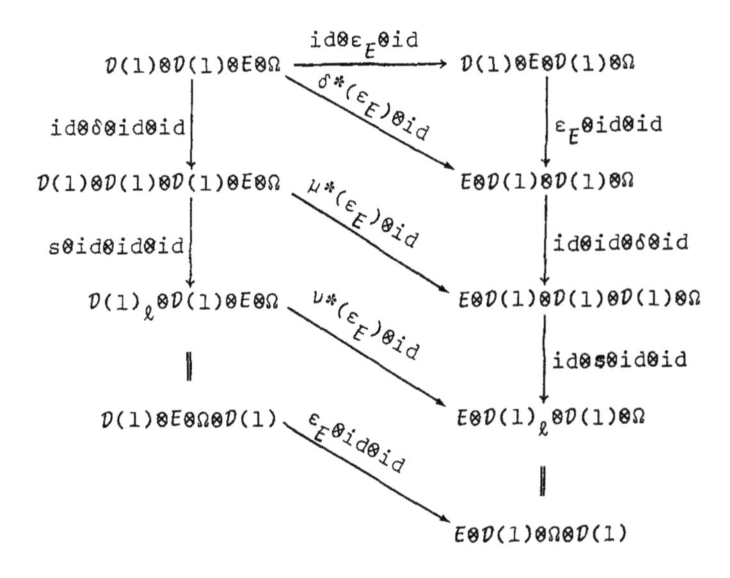

The top triangle is exactly the cocycle condition for ε_E. The slanted arrows come from the morphisms:
δ, $\mu = (id\otimes\delta)\circ\delta$, $\nu = (s\otimes id)\circ\mu$. The commutativity of the three parallelograms is trivial. Since the outer circuit of the diagram is just the following, we have proved that β is horizontal:

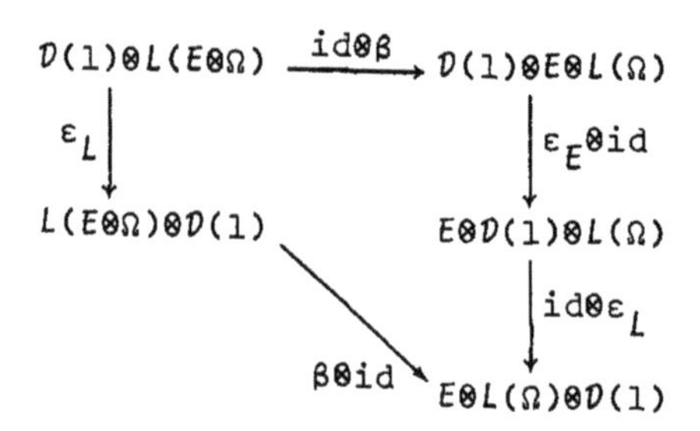

To prove (6.15.2), let $v = \rho(u)$: $\mathcal{D}(1)\otimes E\otimes\Omega \to E\otimes\Omega'$. The assertion is the commutativity of the first diagram below, which expands into the second diagram. Note that again, the essential point is the cocycle condition.

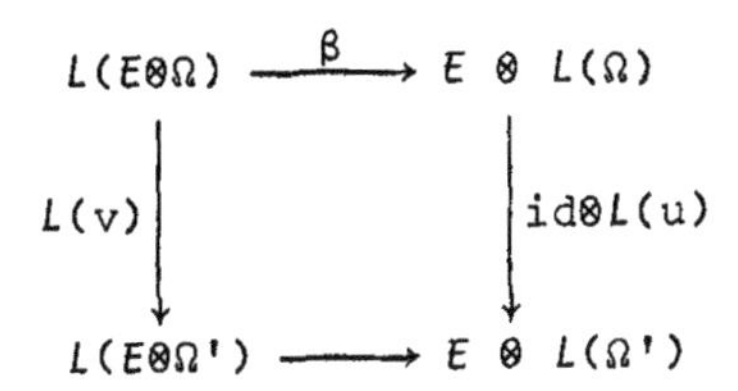
L(E⊗Ω)
β
E ⊗ L(Ω)
L(v)
id⊗L(u)
L(E⊗Ω')
E ⊗ L(Ω')

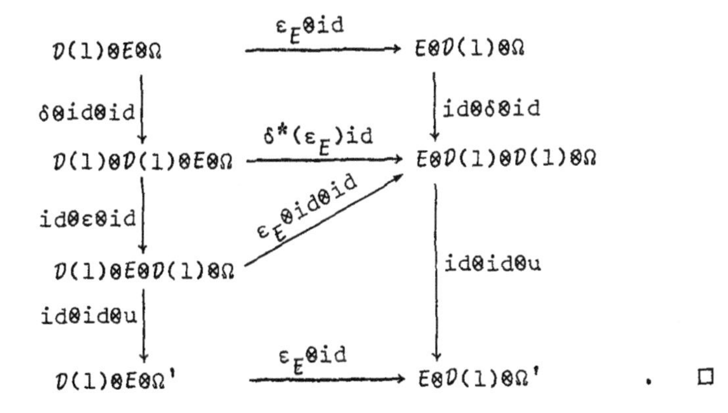
D(1)⊗E⊗Ω
ε_E⊗id
E⊗D(1)⊗Ω
δ⊗id⊗id
id⊗δ⊗id
D(1)⊗D(1)⊗E⊗Ω
δ*(ε_E)id
E⊗D(1)⊗D(1)⊗Ω
id⊗ε⊗id
ε_E⊗id⊗id
D(1)⊗E⊗D(1)⊗Ω
id⊗id⊗u
id⊗id⊗u
D(1)⊗E⊗Ω'
ε_E⊗id
E⊗D(1)⊗Ω'

. □

§7. The Cohomology of a Crystal.

We are now ready to establish the fundamental property of crystalline cohomology, namely, its relation to de Rham cohomology. Our first goal is the following result, from which all the finiteness and base changing properties to come will be deduced:

7.1 Theorem. Suppose $i: X \hookrightarrow Y$ is a closed immersion of S-schemes, with Y/S smooth. Let $\mathcal{E}$ be a $\mathcal{D}_{X,Y}(Y)$-module with HPD stratification, let E be the crystal on X obtained from $\mathcal{E}$ by (6.6), and let $\mathcal{E}\otimes_{\mathcal{D}}\Omega^{\cdot}_{D_{X,Y}(Y)/S}$ be the complex of sheaves on X_{zar} obtained from the connection on $\mathcal{E}$. (Recall that the complex $\mathcal{E}\otimes_{\mathcal{D}}\Omega^{\cdot}_{D/S}$ is the complex $\mathcal{E}\otimes_{\mathcal{O}_Y}\Omega^{\cdot}_{Y/S}$, obtained from the connection on $\mathcal{E}$ as $\mathcal{O}_Y$-module.) Then there is a canonical isomorphism:

(7.1.1) $$H^i(X/S_{cris}, E) \longrightarrow \mathbb{H}^i(X_{zar}, \mathcal{E}\otimes_{\mathcal{D}}\Omega^{\cdot}_{D/S}) .$$

If $u_{X/S}: (X/S)_{cris} \to X_{zar}$ is the canonical projection (5.18), there is a natural isomorphism in the derived category of sheaves of abelian groups on X_{zar}:

(7.1.2) $$\mathbb{R}u_{X/S*}(E) \to \mathcal{E}\otimes_{\mathcal{D}}\Omega^{\cdot}_{D/S} .$$

Proof. Of course, (7.1.2) is a fancy local form of (7.1.1), which in fact it implies, because $\mathbb{R}\Gamma_{cris} = \mathbb{R}\Gamma_{zar} \circ \mathbb{R}u_{X/S*}$. Let us begin by proving (7.1.2) in the special case in which $\mathcal{E}$ arises from an $\mathcal{O}_Y$-module F with HPD stratification, i.e. $\mathcal{E} \cong \mathcal{D}_{X,Y}(Y)\otimes F = \lambda^*F$ (with the HPD stratification of the tensor product). In this case, the Poincaré lemma (6.14) gives us a quasi-isomorphism in $(X/S)_{cris}$: $E \to L(F \otimes_{\mathcal{O}_Y} \Omega^{\cdot}_{Y/S})$. We now use the localization description of L (6.10): Recall that $\tilde{Y} = i^*(Y)$ is represented by $D_{X,\gamma}(Y) \in \mathrm{Cris}(X/S)$, and that we have a commutative diagram:

$$\begin{array}{ccccc} (X/S)_{cris}|_{\tilde{Y}} & \xrightarrow{\varphi} & D_{X,\gamma}(Y)_{zar} & & \\ \downarrow j_{\tilde{Y}} & & \| & \searrow \lambda & \\ (X/S)_{cris} & \xrightarrow{u_{X/S}} & X_{zar} & \hookrightarrow & Y_{zar} \end{array}$$

Now (6.10) tells us that

$L(F \otimes \Omega^q_{Y/S}) \cong j_{\tilde{Y}*}(\lambda \circ \varphi)^*(F \otimes \Omega^q_{X/S}) \cong j_{\tilde{Y}*}\varphi^*(E \otimes \Omega^q_{Y/S})$; note that each of these is acyclic for u_{X/S_*} by (5.27). It follows that there is an isomorphism in the derived category:

$$\mathbb{R}u_{X/S_*}E \xrightarrow{\cong} u_{X/S_*}j_{\tilde{Y}*}\varphi^*(E \otimes_{\mathcal{D}} \Omega^{\cdot}_{D/S}) \cong \varphi_*\varphi^*(E \otimes_{\mathcal{D}} \Omega^{\cdot}_{D/S}) \; .$$

The latter is just the complex of Zariski sheaves $E \otimes_{\mathcal{D}} \Omega^{\cdot}_{D/S}$, so the theorem is proved in our special case. For the general case, we need a lemma:

7.1.2 <u>Lemma.</u> $\mathbb{R}\, i_{cris_*}E \cong i_{cris_*}E$, and the latter is a crystal of $\mathcal{O}_{Y/S}$-modules.

<u>Proof.</u> Since (6.2) tells us that i_{cris_*} is exact, $\mathbb{R}\, i_{cris_*}E \cong i_{cris_*}E$. Suppose $u: T' \to T$ is a morphism in $\mathrm{Cris}(Y/S)$, we must show that $u^*(i_{cris*}(E)_T) \to (i_{cris*}E)_{T'}$ is an isomorphism. Recall from (6.2) that $(i_{cris*}E)_T$ is E_D, where

D is a certain PD envelope of $X \cap T$ in T, whose formation is compatible with change of T . Thus, $u^*(i_{cris_*}E_T) = \mathcal{O}_{T'} \otimes_{\mathcal{D}} E_D \cong \mathcal{D}' \otimes_{\mathcal{D}} E_D \cong E_{D'}$, since E is a crystal on X/S. □

In particular, $E = (i_{cris_*}E)_Y = \lambda_*(E_D)$, with $D = D_{X,\gamma}(Y)$, and $i_{cris_*}(E)$ is the crystal on Y/S defined by E viewed as an HPD stratified $\mathcal{O}_Y$-module. Applying the above special case to (E,∇) on Y/S, we see that there is a canonical isomorphism:

$$\mathbb{R}u_{Y/S_*}\, i_{cris_*}E \cong E \otimes_{\mathcal{O}_Y} \Omega^{\cdot}_{Y/S} \cong E \otimes_{\mathcal{D}} \Omega^{\cdot}_{D/S} \quad .$$

Since $i_{cris_*}E \cong \mathbb{R}i_{cris_*}E$ and $\mathbb{R}u_{Y/S_*}\mathbb{R}i_{cris_*}E = \mathbb{R}i_{zar_*}\mathbb{R}u_{X/S_*}E$, and since i_{zar_*} is exact and $i_{zar}^{-1} \circ i_{zar_*} \cong \mathrm{id}$, the theorem follows. □

Using the filtered Poincaré lemma and (6.10.1), we can prove a more precise result:

7.2 <u>Theorem.</u> Let $D = D_{X,\gamma}(Y)$ and let $F^m(E \otimes \Omega^{\cdot}_{D/S})$ be the subcomplex of $\Omega^{\cdot}_{D/S}$ which in degree q is $J^{[m-q]}E \otimes \Omega^q_{D/S}$ (where J is the ideal of X in D). Then:

$$(7.2.1) \quad \mathbb{R}\Gamma(X/S_{cris}, J^{[m]}_{X/S}E) \xrightarrow{\cong} \mathbb{R}\Gamma(X/S, F^m(E \otimes \Omega^{\cdot}_{D/S})) \quad .$$

$$(7.2.2) \quad \mathbb{R}u_{X/S_*}J^{[m]}_{X/S}E \xrightarrow{\cong} F^m(E \otimes \Omega^{\cdot}_{D/S}) \quad .$$

<u>Proof.</u> The filtered Poincaré lemma (6.14) tells us that we have a quasi-isomorphism:

$$J^{[m]}_{X/S}E \longrightarrow F^m L(E \otimes_{\mathcal{D}} \Omega^{\cdot}_{D/S})$$

Recall that $F^mL(E\otimes\Omega^{\cdot}_{D/S})^q = j_{\tilde{Y}_*}[J^{[m-q]}_{X/S}|_{\tilde{Y}}\ \varphi^*(E\otimes\Omega^q_{D/S})]$ is acyclic for u_{X/S_*} by (5.27.2), and hence applying $\mathbb{R}u_{X/S_*} = u_{X/S_*}$, get $\varphi_*[J^{[m-q]}_{X/S}|_{\tilde{Y}}\ \varphi^*(E\otimes\Omega^q_{D/S})] \cong J^{[m-q]}E\otimes\Omega^q_{D/S}$. The theorem follows. □

7.3 Corollary. There is a natural isomorphism:

$$H^*(X/S_{cris}, \mathcal{O}_{X/S}) \longrightarrow \mathbb{H}^*(X_{zar}, \Omega^{\cdot}_{D/S}) . \quad \square$$

An immediate consequence is the fact that crystalline cohomology captures the de Rham cohomology of a lifting, if it exists.

7.4 Corollary. Suppose X/S is smooth, $S_0 \hookrightarrow S$ is defined by a sub PD ideal of I, and $X_0 = S_0 \underset{S}{\times} X$. Then there is a natural quasi-isomorphism:

$$H^*(X_0/S_{cris}, \mathcal{O}_{X_0/S}) \longrightarrow \mathbb{H}^*(X_{zar}, \Omega^{\cdot}_{X/S}) .$$

Proof. Just recall that we showed in (5.17) that $H^*(X_0/S_{cris}, \mathcal{O}_{X_0/S}) \cong H^*(X/S_{cris}, \mathcal{O}_{X/S})$ and use (7.1) with X = Y. □

7.5 Remark. Because the isomorphism of (7.3) is natural, we can use it to compute the crystalline cohomology of maps, as well as spaces. If $f: X' \to X$ is a map over a PD morphism $\varphi: (S', I', \gamma') \to (S, I, \gamma)$, if X' and X can be embedded in schemes which are smooth over S' and S, and if X/S is separated, then in fact we can find such embeddings $X' \hookrightarrow Y'$, $X \hookrightarrow Y$, and a map $F: Y' \longrightarrow Y$ which induces f. We obtain a commutative diagram:

$$\begin{array}{ccc} H^*(X/S_{\text{cris}}, \mathcal{O}_{X/S}) & \xleftrightarrow{\cong} & \mathbb{H}^*(X_{\text{zar}}, \Omega^\cdot_{\mathcal{D}_{X,\gamma}(Y)/S}) \\ \downarrow f^* & & \downarrow F^* \\ H^*(X'/S'_{\text{cris}}, \mathcal{O}_{X'/S'}) & \xleftrightarrow{\cong} & \mathbb{H}^*(X'_{\text{zar}}, \Omega^\cdot_{\mathcal{D}_{X',\gamma}(Y')/S'}) \end{array}$$

It follows that F^* depends only on f, and not on Y', Y, or F.

A crystal E of $\mathcal{O}_{X/S}$-modules on Cris(X/S) is said to be "quasi-coherent" iff each $E_{(U,T,\delta)}$ is a quasi-coherent sheaf of $\mathcal{O}_T$-modules on T. Our next task is the establishment of finiteness and base changing properties of the cohomology of quasi-coherent crystals.

7.6 Theorem. *Suppose S is quasi-compact, $f: X \to (S,I,\gamma)$ is quasi-compact and quasi-separated, and $f_{X/S}: (X/S)_{\text{cris}} \to S$ is the composite $f \circ u_{X/S}$. Then for any quasi-coherent crystal E in $(X/S)_{\text{cris}}$, $R^i f_{X/S*} E$ is quasi-coherent on S. Moreover there exists an r such that $R^i f_{X/S*} E = 0$ for $i > r$ and all E.*

Proof. First suppose X can be embedded in a scheme Y which is smooth, quasi-compact, and quasi-separated over S. Then we get from (7.1) a natural isomorphism in the derived category: $\mathbb{R}f_{X/S*} E \cong \mathbb{R}f_*(E_{D_{X,\gamma}(Y)} \otimes_{\mathcal{O}_Y} \Omega^\cdot_{Y/S})$. Since $E_{D_{X,\gamma}(Y)} \otimes_{\mathcal{O}} \Omega^\cdot_{Y/S}$ is a bounded complex of quasi-coherent $\mathcal{O}_Y$-modules, the theorem for our X follows from standard considerations.

We can reduce to the above case by using Cech cohomology. Let $U = \{U_i\}$ be a finite affine open covering of X, and for each $\nu \geq 0$, let $U_{(\nu)} = \coprod_{i_0 < \dots i_\nu} U_{i_0} \cap \dots U_{i_\nu}$. If F is a sheaf of

$\mathcal{O}_{X/S}$-modules on Cris(X/S), let $C^{\nu}(F) = j_{(\nu)cris_*} j_{(\nu)cris}{}^*(F)$, where $j_{(\nu)}: U_{(\nu)} \to X$ is the natural map. Then we can form a complex $C^{\cdot}(F)$ in a natural way, which is easily seen to be a resolution of F. Now let $E \to E^{\cdot}$ be a resolution of E by injective sheaves of $\mathcal{O}_{X/S}$-modules. One sees easily that the total complex $C^{\cdot}(E^{\cdot})$ is a resolution of E by sheaves which are acyclic for $f_{X/S*}$. Thus the cohomology of the double complex $f_{X/S_*}(C^{\cdot}(E^{\cdot}))$ is the cohomology of $\mathbb{R}f_{X/S*}E$. The spectral sequence of the double complex $E^{pq} = f_{X/S_*}C^p(E^q)$ therefore converges to $R^i f_{X/S_*}E$, so we need only prove that each E_1^{pq} is quasi-coherent and that $E_\infty^{pq} = 0$ for p and q large — as is already clear for p. Now E_1^{pq} is the cohomology of the complex $E_0^p = f_{X/S_*} j_{(p)cris_*} j^*_{(p)cris}(E^{\cdot})$ and since the complex $j^*_{(p)cris}(E^{\cdot})$ is a resolution of $j^*_{(p)cris}(E)$ by sheaves which are acyclic for $f_{U_{(p)}/S*} = f_{X/S*} \circ j_{(p)*}$, $E_1^{pq} = R^q f_{U_{(p)}/S_*} j^*_{(p)}(E))$. Thus, we have reduced the problem to the spaces $U_{(p)}$, since $j^*_{(p)}(E)$ is a quasi-coherent crystal on $U_{(p)}$. Clearly it suffices to prove the result for the sets $U_{i_0} \cap \ldots \cap U_{i_p}$.

Since X/S is quasi-separated, sets of the above form are not necessarily affine, but <u>are</u> quasi-compact. They are also obviously quasi-affine, and hence separated. In other words, we are reduced to proving the theorem for X quasi-compact and separated. Repeat the above Cech procedure with an affine covering. This time one gets an affine covering with affine intersections, so the result follows from the affine case. □

Of course it would be tedious to attempt to write down a specific r which worked in the above theorem. However, it is important to notice that the process of embedding pieces of X in smooth schemes is compatible with base change, so that (7.6) really gives an r which works not just for X/S, but also for any X'/S' obtained from base change via a PD morphism $(S',I',\gamma') \to (S,I,\gamma)$.

In order to study the base changing properties of the cohomology of a crystal, we first need a fancy form of the adjunction formula in the derived category. This holds for any morphism of ringed topoi, as explained in [SGA4 XVII], where it is called "trivial duality". Let us give the reader a rough sketch:

Recall first that if $f\colon (T',A') \to (T,A)$ is a morphism of ringed topoi, and if $F^\cdot$ is a complex of A'-modules which is bounded below, then it is always possible to find a quasi-isomorphism $F^\cdot \to I^\cdot$, where $I^\cdot$ is a complex of injective A'-modules in T' and the hyperderived functor $\mathbb{R}f_*F^\cdot$ is, by definition the image of $f_*I^\cdot$ in the derived category of A-modules in T.

We need a similar construction for f^*. Since there do not exist, in general, enough projectives in the category of A-modules in a topos, the construction is not so standard, so we will indicate a few more details. First observe that there is a functor L^0 which associates to any A-module E a flat A-module $L^0(E)$ and an epimorphism $L^0(E) \to E$. Namely $L^0(E)$ is the sheafification of the presheaf which takes any U to *the free A-module with basis* $E(U) - \{0\}$. (*Leaving out* 0 insures us that L^0 takes zero maps to zero maps.) It follows that there is a functorial complex $L^\cdot(E)$ of flat A-modules with $H^i(L^\bullet(E)) = 0$ if $i \neq 0$, and $H^0(L^\bullet(E))$ naturally isomorphic to E. Now if $E^\bullet$ is a complex of A-modules, $L^\bullet(E^\bullet)$ is a double complex of flat A-modules, and the associated simple complex $L^\bullet(E^\bullet)$ is easily seen to be naturally quasi-isomorphic to E if $E^\bullet$ is bounded above. If $E^\bullet$ and $F^\bullet$ are two such complexes, the natural map $L^\bullet(E^\bullet) \otimes L^\bullet(F^\bullet) \to L^\bullet(E^\bullet \otimes F^\bullet)$ is hence a quasi-isomorphism, and since $L^\bullet$ is also compatible with shifts, one see that if $C^\bullet(u)$ is the mapping cone of a morphism of complexes $u: E^\bullet \to F^\bullet$, then there is a natural quasi-isomorphism $C(L^\bullet(u)) \to L^\bullet(C^\bullet(u))$. (That is, $L^\bullet$ preserves triangles.) We now define $\mathbb{L}f^*(E^\bullet)$ to be the class of $f^*L^\bullet(E^\bullet)$ in the derived category of A-modules. Since the modules comprising $L^\bullet(E^\bullet)$ are flat, $\mathbb{L}f^* = f^*L^\bullet$ takes acyclic complexes to acyclic complexes, hence quasi-isomorphism to quasi-isomorphisms, (and triangles to triangles). Note that $\mathbb{L}f^*E^\bullet$ is only defined on complexes which are bounded above.

Finally, if $M^\bullet$ is bounded above and $N^\bullet$ is bounded below, $\mathbb{R}\,\mathrm{Hom}^\bullet_A[M^\bullet,N^\bullet]$ is defined to be the class of the complex $\mathrm{Hom}^\bullet_A[M^\bullet,I^\bullet]$, where $I^\bullet$ is an injective resolution of $N^\bullet$. (Recall that this is the complex which in degree k is $\prod_i \mathrm{Hom}[M^i,I^{i+k}]$, with the usual boundary maps.) Note that since $M^\bullet$ is bounded above and $I^\bullet$ is bounded below, the product $\prod_i$ is really only a finite one.

7.7 Proposition. (Adjunction formula). If $E^\bullet$ is a complex of A-modules bounded above, if $F^\bullet$ is a complex of A'-modules bounded below, and if $f: (T',A') \to (T,A)$ is a morphism, there is a canonical, transitive, functorial isomorphism:

$$\mathrm{Ad}_f\colon\ \mathbb{R}\,\mathrm{Hom}^{\bullet}_{A'}(\mathbb{L}f^{*}E^{\bullet},F^{\bullet}) \to \mathbb{R}\,\mathrm{Hom}^{\bullet}_{A}(E^{\bullet},\ \mathbb{R}f_{*}F^{\bullet})$$

in the derived category of abelian groups.

<u>Proof.</u> Let $L^{\bullet} \to E^{\bullet}$ be a flat resolution, and let $F^{\bullet} \to I^{\bullet}$ and $f_{*}I^{\bullet} \to J^{\bullet}$ be injective resolutions. Then $\mathbb{L}f^{*}E^{\bullet}$ is $f^{*}L^{\bullet}$, $\mathbb{R}\,\mathrm{Hom}^{\bullet}_{A'}(\mathbb{L}f^{*}E^{\bullet},F^{\bullet})$ is $\mathrm{Hom}^{\bullet}_{A'}(f^{*}L^{\bullet},I^{\bullet})$, and $\mathbb{R}\,\mathrm{Hom}^{\bullet}_{A}(E^{\bullet},\mathbb{R}f_{*}F^{\bullet})$ is $\mathrm{Hom}^{\bullet}_{A}(E^{\bullet},J^{\bullet})$. Using the adjointness of f^{*} and f_{*}, we obtain a natural map:

$$\mathrm{Ad}_f\colon\ \mathbb{R}\,\mathrm{Hom}^{\bullet}_{A'}(\mathbb{L}f^{*}E^{\bullet},F^{\bullet}) = \mathrm{Hom}^{\bullet}_{A'}(f^{*}L^{\bullet},I^{\bullet}) \cong \mathrm{Hom}^{\bullet}_{A}(L^{\bullet},\ f_{*}I^{\bullet}) \longrightarrow$$

$$\longrightarrow \mathrm{Hom}^{\bullet}_{A}(L^{\bullet},J^{\bullet}) \xleftarrow{\ \sim\ } \mathrm{Hom}^{\bullet}_{A}(E^{\bullet},J^{\bullet}) = \mathbb{R}\,\mathrm{Hom}^{\bullet}_{A}(E^{\bullet},\mathbb{R}f_{*}F^{\bullet})\ .$$

(The fact that the arrow in the second line is a quasi-isomorphism depends on the injectivity of the modules J^{q}.)

This gives us the desired arrow. We shall not verify, or even list, all the compatibilities it satisfies. However we shall use them freely, even in our sketch of the proof of the fact that it is an isomorphism. Clearly the problem is in the arrow $\mathrm{Hom}^{\bullet}_{A}(L^{\bullet},f_{*}I^{\bullet}) \to \mathrm{Hom}^{\bullet}_{A}(L^{\bullet},J^{\bullet})$.

We begin by factoring f into two morphisms: $(T',A') \xrightarrow{g} (T',f^{-1}(A)) \xrightarrow{h} (T,A)$ (the reader can, I hope, easily imagine the meaning of the middle term). By appealing to transitivity of Ad_f, we reduce the problem to g and h. Now the functor h^{*} is exact, so its adjoint h_{*} takes injectives to injectives, so the problem is trivial. We have thus reduced the problem to maps of the form $f\colon (T,B) \to (T,A)$.

To prove that the arrow $\mathrm{Hom}^{\bullet}_A(L^{\bullet}, f_* I^{\bullet}) \to \mathrm{Hom}^{\bullet}_A(L^{\bullet}, J^{\bullet})$ is a quasi-isomorphism, we have to show that the complex $f_* I^{\bullet}$ comprises modules which are acyclic for $\mathrm{Hom}^{\bullet}_A(L^{\bullet}, \ )$. In other words, we have to prove that if L is A-flat and I is B-injective, then $\mathrm{Ext}^i_A(L, f_* I) = 0$ for $i > 0$.

Choose an injective A-module J containing $f_* I$; then $I \subseteq \mathcal{H}om_A(f_* B, J)$ in a natural way, and since I is injective, it is in fact a direct summand, and it is enough to prove the statement with this $\mathcal{H}om$ in place of I .

Let $P^{\bullet}$ be a resolution of $f_* B$ by flat A-modules. The flatness of the P's and the adjointness of $\otimes$ and $\mathcal{H}om$ show that $\mathcal{H}om_A(P^{\bullet}, J)$ is a complex of injective A-modules, in fact an injective resolution of $\mathcal{H}om_A(f_* B, J)$. Thus we have only to calculate the cohomology of the complex $\mathrm{Hom}_A(L, \mathcal{H}om_A(P^{\bullet}, J))$. But this is just $\mathrm{Hom}_A(L \otimes P^{\bullet}, J)$, and since L is flat and J is injective, it's just a resolution of $\mathrm{Hom}_A(L \otimes B, J)$, and in particular is acyclic. □

Perhaps we should remark that an A-module E in ringed topos $(\mathcal{T}, A)$ is defined to be flat iff $E \otimes_A$ is exact. It is not hard to prove that a sheaf E of $\mathcal{O}_{X/S}$-modules in $(X/S)_{\mathrm{cris}}$ turns out to be flat iff each E_T is a flat $\mathcal{O}_T$-module [Berthelot III 3.5.2], but we may as well take this as our definition.

We are now ready to study the base changing properties of the cohomology of a flat crystal. Suppose S is quasi-compact, $f: X \to (S, I, \gamma)$ and $f': X' \to (S', I', \gamma')$ are quasi-compact and quasi-separated, and $g: X' \to X$ covers a PD morphism $u: (S', I', \gamma') \to (S, I, \gamma)$.

Let E be a quasi-coherent, flat crystal on X. Then (7.6) tells us that the complex $\mathbb{R}f_{X/S_*}E$ is actually bounded, so $\mathbb{L}u^*\mathbb{R}f_{X/S_*}E$ is defined. On the other hand, $E' = g^*E$ is a crystal on X', and $\mathbb{R}f'_{X'/S'_*}E'$ is a bounded complex on S', which is reasonable to compare with $\mathbb{L}u^*\mathbb{R}f_{X/S_*}E$.

In fact it is easy to obtain a map (in the derived category) $\mathbb{L}u^*\mathbb{R}f_{X/S_*}E \to \mathbb{R}f'_{X'/S'_*}E'$, i.e. an element of H^0 of the complex $\mathbb{R}\,\mathrm{Hom}_{S'}(\mathbb{L}u^*\mathbb{R}f_{X/S_*}E, \mathbb{R}f'_{X'/S'_*}E')$. This can be done from the adjunction formula in several ways, here is one: The adjunction formula for g gives an isomorphism:
$\mathbb{R}\,\mathrm{Hom}^\cdot[\mathbb{L}g^*_{cris}E, E'] \to \mathbb{R}\,\mathrm{Hom}^\cdot[E, \mathbb{R}g_{cris_*}E']$, and since E is flat, $\mathbb{L}g^*_{cris}E = E'$ and we obtain (from the identity map $id_{E'}$) a morphism in the derived category: $E \to \mathbb{R}g_{cris*}E'$. Applying the functor $\mathbb{R}f_{X/S_*}$, we obtain an element of

$$\mathbb{R}\,\mathrm{Hom}_S(\mathbb{R}f_{X/S_*}E, \mathbb{R}f_{X/S_*}\mathbb{R}g_{cris*}E') \cong \mathbb{R}\,\mathrm{Hom}_S(\mathbb{R}f_{X/S_*}E, \mathbb{R}u_*\mathbb{R}f'_{X'/S'*}E').$$

By the adjunction formula for u, this in turn is isomorphic to $\mathbb{R}\,\mathrm{Hom}_S(\mathbb{L}u^*\mathbb{R}f_{X/S_*}E, \mathbb{R}f'_{X'/S'*}E')$, which is where we wanted to be.

Some additional hypotheses will allow us to assert that the base changing map above is actually an isomorphism. We suppose that the morphisms f and f' fit into the diagram:

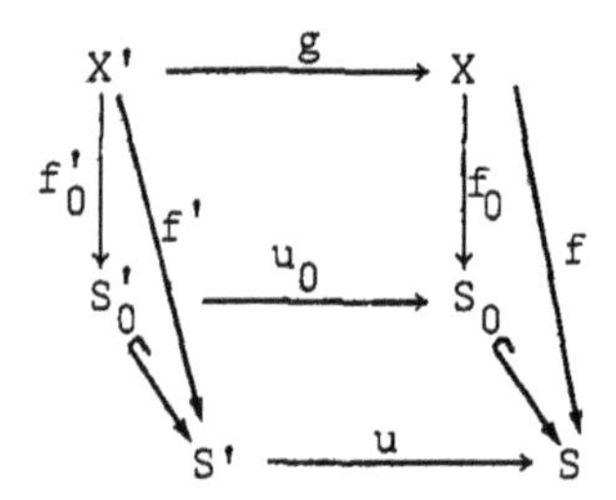

7.8 Theorem. Suppose that in the above diagram, S is quasi-compact, f_Q is smooth, quasi-compact, and quasi-separated, u is a PD morphism $(S',I',\gamma') \to (S,I,\gamma)$, and $S_0 \subseteq S$ and $S_0' \subseteq S'$ are defined by sub PD ideals of I and I', respectively. Suppose further that $X' = X \times_{S_0} S_0'$ and that E is a flat, quasicoherent crystal on X. Then the base changing arrow is an isomorphism:

$$Lu^* \mathbb{R} f_{X/S_*} E \longrightarrow \mathbb{R} f'_{X'/S'_*} g^*_{cris} E \ .$$

Proof. We begin with a very special case, assuming that X lifts to a smooth affine scheme Y over S. Then $Y' = Y \times_S S'$ lifts X', and we can write everything down explicitly: Since $S_0 \subseteq S$ is defined by a sub PD ideal, $D_{X,\gamma}(Y) = Y$, and the crystal E corresponds to an $\mathcal{O}_Y$-module $E = E_{D_{X,\gamma}(Y)}$ with integrable quasi-nilpotent connection. Since E is flat, each E_T is flat over $\mathcal{O}_T$, so E is a flat $\mathcal{O}_Y$-module. We know from (7.1.2) that $\mathbb{R} f_{X/S*} E \cong \mathbb{R} f_* \mathbb{R} u_{X/S_*} E \cong \mathbb{R} f_* (E \otimes \Omega^{\cdot}_{Y/S})$, and since f is affine, this is just the complex $f_*(E \otimes \Omega^{\cdot}_{Y/S})$. On the other hand, the crystal $g^*_{cris} E = E'$ corresponds to the $\mathcal{O}_{Y'}$-module $E' = pr^* E$, where $pr: Y' \to Y$ is the natural map, and we have $\mathbb{R} f'_{X'/S'*} E' \cong f'_*(E' \otimes \Omega^{\cdot}_{Y'/S'})$. Since $Y' = Y \times_S S'$, we have a natural isomorphism of complexes of $\mathcal{O}_{S'}$-modules: $u^*(f_*(E \otimes \Omega^{\cdot}_{Y/S})) \longrightarrow f'_*(E' \otimes \Omega^{\cdot}_{Y'/S'})$ — this is just the base changing theorem for affine morphines in the Zariski topology, together with the fact that formation of $\Omega^{\cdot}$ is compatible with base change. Of course, this arrow is the same as our fancy looking base changing arrow, and the theorem is proved in the special case.

To deduce the general case it is necessary, unfortunately, to resort to the technique of cohomological descent. This is because it is not clear a priori that the adjunction map is compatible with Cech resolution. In order to overcome this difficulty, we have to provide the Cech resolution with a geometric meaning – i.e. we have to express it in terms of topoi. We shall give here only a rough sketch and refer to [SGA4 V^{bis}] for details.

The idea of cohomological descent is to replace the topos $(X/S)_{cris}$ by a topos $(X^{\bullet}/S)$ which stands for a finite covering $\{U_i\}$ of X together with all the gluing data. Thus if $i = (i_0,\ldots,i_k)$ is a multi-index with $i_0 < i_1 < \ldots i_k$, let $U_i = U_{i_0} \cap \ldots U_{i_k}$ and consider the simplicial scheme:

$$\ldots \overrightarrow{\overrightarrow{\rightarrow}} \coprod_{|i|=2} U_i \rightrightarrows \coprod_{|i|=1} U_i \quad .$$

We construct a topos $(X^{\bullet}/S)_{cris}$ whose "sheaves" are collection of sheaves $\{F_i$ on each $U_i\}$ together with compatible maps between the F_i's covering the inclusions $U_i \to U_j$. There is an obvious morphism of topoi: $\pi\colon (X^{\bullet}/S)_{cris} \to (X/S)_{cris}$, deduced from the inclusions $U \longrightarrow X$. The key fact of cohomological descent, which we shall use without proof, is that for any abelian sheaf E, the natural map, $E \longrightarrow \mathbb{R}\pi_*\pi^*E$ is an isomorphism.

Using this, we can reduce our proof to the topos $(X^{\bullet}/S)_{cris}$: Construct a topos $(X'^{\bullet}/S')_{cris}$ in a similar manner, using the covering $\{g^{-1}(U_i)\}$ of X'. If $f_{X^{\bullet}/S} = f_{X/S} \circ \pi$, etc., we get a commutative diagram:

$$\begin{array}{ccc}
\mathbb{L}u^* \, \mathbb{R}f_{X^\bullet/S_*}\pi^* E & \longrightarrow & \mathbb{R}f'_{X'^\bullet/S'_*}(g^{\bullet *}_{cris}\pi^* E) \\
\downarrow \cong & & \downarrow \cong \\
\mathbb{L}u^* \, \mathbb{R}f_{X/S_*}\mathbb{R}\pi_*\pi^* E & \longrightarrow & \mathbb{R}f'_{X^\bullet/S'_*}\, \mathbb{R}\pi'_*\pi'^* \, g^*_{cris}E \\
\downarrow \cong & & \downarrow \cong \\
\mathbb{L}u^* \, \mathbb{R}f_{X/S_*}E & \longrightarrow & \mathbb{R}f'_{X'/S'_*}g^*_{cris}E
\end{array}$$

Thus, it suffices to prove that the top arrow is an isomorphism.

We would like to reduce to the individual opens U_i in the open covering. To do this, we have to construct a stupid topos $S^\bullet$ whose sheaves are families of sheaves on $S_{\underline{zar}}$ indexed by the same indices as before. The point is that $f_{X^\bullet/S}$ factors through an obvious map $\omega\colon S^\bullet \to S$: $f_{X^\bullet/S} = \omega \circ f_{X^\bullet/S^\bullet}$, and to prove the base changing theorem for $f_{X^\bullet/S}$ it suffices to do it for ω and $f_{X^\bullet/S^\bullet}$. Since it is easy for ω, we are reduced to the morphism $f_{X^\bullet/S^\bullet}$, and the amounts to checking it for each map $U_i \to S$, as is easy to see.

We can now easily complete the proof of the theorem. So far, we know it if either:

(i) X lifts to a smooth affine scheme Y over S, or

(ii) X has a finite open covering $\{U_i\}$ such that the theorem holds for any intersection $U_{i_0} \cap \ldots U_{i_k}$.

Suppose first that X satisfies (i) and $U \subset X$ is open and quasi-compact. Then it is clear that U can be covered by finitely many special affine sets U_i which satisfy (i), and it is also clear that the intersections $U_{i_0} \cap \ldots \cap U_{i_k}$ also satisfy (i). Thus the theorem is true for U. Finally, suppose X/S_0 is smooth, separated, and quasi-compact. If $x \in X$, it is easy to see, by choosing local coordinates at x, that x has an open neighborhood U_x satisfying (i) – the local coordinates mean that we have to lift an étale $\mathcal{O}_{S_0}[t_1 \ldots t_n]$-scheme to $\mathcal{O}_S[t_1 \ldots t_n]$, which is possible because $S_0 \subseteq S$ is a nil immersion – c.f. [EGA IV 18.1.2]. Since X is quasi-compact, finitely many of these neighborhoods will do, and since X is quasi-separated, any intersection U of them is quasi-compact. Hence the covering we get satisfies (ii), so the theorem is true for X. □

7.9 Corollary. Let S be quasi-compact, let $f: X \to S_0$ be smooth, quasi-compact, and quasi-separated, with $S_0 \subseteq (S, I, \gamma)$ defined by a sub PD ideal of I. Let E be a flat crystal on X/S, and let $E = E_{(X,X)}$, with ∇ its associated integrable connection (relative to S_0). Then there is an isomorphism:

$$(\mathbb{R}f_{X/S_*}E) \otimes^{\mathbf{L}} \mathcal{O}_{S_0} \longrightarrow \mathbb{R}f_*(E \otimes \Omega^{\cdot}_{X/S_0})$$

Proof. The base changing theorem tells us that $\mathbb{R}f_{X/S_*}E \otimes^{\mathbf{L}} \mathcal{O}_{S_0}$ is just $\mathbb{R}f_{X/S_0*}(E|(X/S_0)_{cris})$, and we know that the latter is $\mathbb{R}f_*(E \otimes \Omega^{\cdot}_{X/S_0})$. □

7.10 Remark. Note that in the equicharacteristic case (i.e. $p\mathcal{O}_S = 0$), if f is proper and flat, it is still not known whether or not each $R^i f_{X/S_*} \mathcal{O}_{X/S}$ commutes with base change individually — i.e. whether or not each is locally free.

7.11 Corollary. Suppose $f: X \to Y$ is smooth, quasi-compact, and quasi-separated, and Y is quasi-compact. Then if E is a flat quasi-coherent crystal on X, $\mathbb{R}f_{\text{cris}_*}E$ is a crystal in the derived category of $\mathcal{O}_{Y/S}$-modules.

Proof. Suppose $T = (U,T,\delta)$ is an object of Cris(Y/S). Let $W = f^{-1}(U)$. It is easy to see that there is a natural equivalence of categories $(X/S)_{\text{cris}}\big|_{f^{-1}(T)} \cong (W/T)_{\text{cris}}$, and using this, one can identify $(\mathbb{R}f_{\text{cris}_*}E)_T$ with $\mathbb{R}f_{W/T_*}(E|_{(W/T)_{\text{cris}}})$. Now if $v: (U',T',\delta') \to (U,T,\delta)$ is a morphism in $(Y/S)_{\text{cris}}$, $W' = f^{-1}(U') \cong W \times_T T'$, and since $U \subseteq T$ and $U' \subseteq T'$ are defined by PD ideals, (7.8) tells us that the map

$$\mathbb{L}v^*\mathbb{R}\, f_{W/T_*}(E|_{(W/T)_{\text{cris}}}) \longrightarrow \mathbb{R}\, f_{W'/T'}(E|_{(W'/T')_{\text{cris}}})$$

is an isomorphism, as desired. □

This result is a crystalline version of the Gauss-Manin connection. We can make this explicit when Y/S is smooth. In this case the maps $p_i: D_{Y/S}(1) \to Y$ are flat, so $\mathbb{L}p_i = p_i^*$, and we see from (7.11) that there is an isomorphism:

$$p_2^*[(\mathbb{R}^i f_{\text{cris}*}E)_Y] \longrightarrow p_1^*[(\mathbb{R}^i f_{\text{cris}*}E)_Y] .$$

Since $f: X \longrightarrow Y$ is smooth $(\mathbb{R}^i f_{\text{cris}*}E)_Y$ is $\mathbb{R}^i f_*(E_X \otimes \Omega^\cdot_{X/Y})$, and we have exactly an integrable, quasi-nilpotent connection on $\mathbb{R}^i f_*(E_X \otimes \Omega^\cdot_{X/Y})$ — the Gauss-Manin connection.

Here is a strong form of the base changing theorem for a smooth map:

7.12 Corollary. Suppose, in the diagram below, that f is quasi-separated and smooth, that Y is quasi-compact, and that $X' = X\times_Y Y'$. Then if E is a flat quasi-coherent crystal on E, the base changing map:

$$\mathbb{L}h^*_{cris}\,\mathbb{R}f_{cris*}E \longrightarrow \mathbb{R}f'_{cris*}g_{cris}{}^*E$$

is an isomorphism.

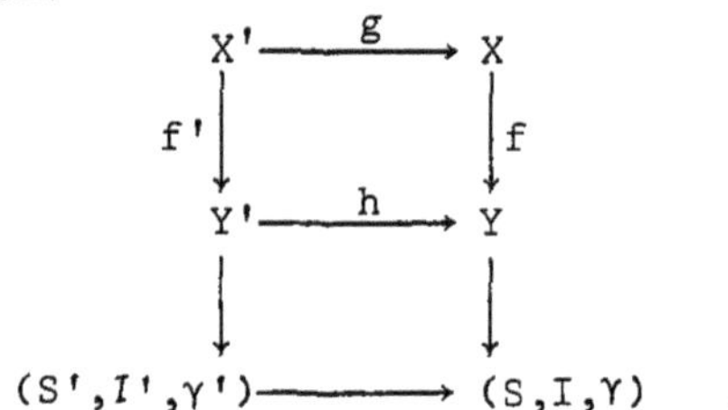

Proof. Suppose $(U',T',\delta') \in \mathrm{Cris}(Y'/S')$, $(U,T,\delta) \in \mathrm{Cris}(Y/S)$, and $v\colon T' \to T$ is a PD morphism. Thanks to (7.11) and Exercise (6.5), we can describe $(\mathbb{L}h^*_{cris}\mathbb{R}f_{cris*}E)_{T'}$ as $\mathbb{L}v^*(\mathbb{R}f_{cris*}E)_T$. But now if we identify $(\mathbb{R}f_{cris*}E)_T$ with $\mathbb{R}f_{X/T*}E$ and $\mathbb{R}f'_{cris*}(g_{cris}{}^*E)_{T'}$ with $\mathbb{R}f_{X'/T'*}(g_{cris}{}^*E)$, we see that the arrow $\mathbb{L}v^*(\mathbb{R}f_{cris*}E)_T \longrightarrow (\mathbb{R}f'_{cris*}(g_{cris}{}^*E))_{T'}$ is an isomorphism, as desired. □

The reader may have already noticed an important consequence of the assertion (7.8) that the arrow: $\mathbb{L}u^*\mathbb{R}_{X/S*}E \to \mathbb{R}f'_{X/S*}g_{cris}{}^*E$ is a quasi-isomorphism: the target is a bounded complex by (7.6), hence $H^i(\mathbb{L}u^*\mathbb{R}f_{X/S*}E) = 0$ for almost all i. This will enable us to prove the following finiteness results for $\mathbb{R}f_{X/S*}E$:

7.13 <u>Corollary.</u> Let $f: X \longrightarrow (S,I,\gamma)$ satisfy the hypotheses of (7.8), and let E be a flat quasi-coherent crystal on Cris(X/S). Then $\mathbb{R}f_{X/S*}E$ has finite tor-dimension, i.e. is isomorphic in the derived category to a bounded complex of flat $\mathcal{O}_S$-modules.

<u>Proof.</u> I claim first of all that there exists an integer n such that for all $\mathcal{O}_S$-modules M, $\mathbb{R}f_{X/S*}E \overset{\mathbb{L}}{\otimes} M$ is acyclic except in degrees within $(0,n)$. Since S is quasi-compact, we can assume S is affine. First suppose that M is quasi-coherent, and consider the $\mathcal{O}_S$-algebra $\mathcal{O}_S \oplus M$, $S' = \mathit{Spec}_S(\mathcal{O}_S \oplus M)$. The ideal I' generated by M and I in $\mathcal{O}_{S'}$ has a PD structure γ', and there is a PD morphism $u: (S',I',\gamma') \to (S,I,\gamma)$. Now the base changing theorem tells us that $\mathbb{R}f_{X/S*}(E) \overset{\mathbb{L}}{\otimes} \mathcal{O}_{S'} \cong \mathbb{R}f'_{X'/S'}(E')$, a complex which we can bound (independent of S') by (7.6). But $\mathbb{R}f_{X/S*}(E)\overset{\mathbb{L}}{\otimes}M$ is a direct summand of $Rf_{X/S*}(E)\overset{\mathbb{L}}{\otimes}\mathcal{O}_{S'}$, and hence it too is uniformly bounded.

The above argument works only for quasi-coherent M, but if M is arbitrary, it implies that the stalks of $\mathbb{R}f_{X/S*}(E)\overset{\mathbb{L}}{\otimes}M$ are uniformly bounded, hence so is $\mathbb{R}f_{X/S*}(E)\overset{\mathbb{L}}{\otimes}M$.

Because $\mathbb{R}f_{X/S*}(E)$ is bounded above, we can find a complex $L^\cdot$ of flat $\mathcal{O}_S$-modules representing it, still bounded above. For each q, let:

$$A^q = L^q \text{ if } q > 0, \quad L^0/d(L^{-1}) \text{ if } q = 0, \quad 0 \text{ if } q < 0.$$

$$B^q = L^q \text{ if } q \le 0, \quad 0 \text{ if } q > n.$$

Then the obvious maps $L^\cdot \to A^\cdot$ and $B^\cdot \to A^0$ are quasi-isomorphisms. Since the B^q's are flat, we have $\mathcal{T}or_i(A^0,M) \cong H^{-i}(B^\cdot \otimes M) \cong H^{-i}(L^\cdot \otimes M) \cong H^{-i}(\mathbb{R}f_{X/S_*} E \overset{\mathbb{L}}{\otimes} M) = 0$ for any M and any $i > 0$. Thus, A^0 is flat, and hence $A^\cdot$ is a bounded complex of flat $\mathcal{O}_S$-modules quasi-isomorphic to $L^\cdot$. This proves the theorem. □

We can now prove the fundamental finiteness property of the cohomology of a crystal. The key concept is that of a perfect complex. This notion is defined and studied in [SGA 6 I] for nonnoetherian ringed topoi. We will content ourselves with a much more prosaic situation:

7.14 <u>Definition.</u> Let A be a noetherian ring. A complex $K^\cdot$ of A-modules is called "strictly perfect" iff it is bounded and if each K^q is finitely generated and projective.

7.15 <u>Lemma.</u> A complex $K^\cdot$ of A-modules is quasi-isomorphic to a strictly perfect complex iff it has finite tor-dimension and finitely generated cohomology.

<u>Proof.</u> One constructs a complex $P^\cdot$ of finitely generated projectives and a quasi-isomorphism $P^\cdot \to K^\cdot$, inductively, in the usual manner. Then as in the proof of the previous result, if $K^\cdot$ is acyclic in degrees $< m$, one has a quasi-isomorphism $P^\cdot \to A^\cdot$, with $A^m = P^m/d(P^{m-1})$, $A^n = 0$ if $n < m$, and $A^n = P^n$ if $n > m$. Then A^m is flat and finitely generated, hence projective. □

7.16 <u>Theorem.</u> Suppose $f:X \longrightarrow S_0$ is a smooth proper map, $S_0 \hookrightarrow S$ is defined by a sub PD ideal of I, and S is noetherian. If E is a crystal of locally free, finite rank $\mathcal{O}_{X/S}$-modules, $\mathbb{R}f_{X/S*}E$ is a perfect complex of $\mathcal{O}_S$-modules — i.e. is, locally on S, quasi-isomorphic to a strictly perfect complex.

<u>Proof.</u> Since the assertion is local on S, we may assume S is affine. Moreover, by the quasi-coherence of crystalline cohomology (7.6), we may work over $A = \Gamma(S, \mathcal{O}_S)$ instead of S. Then (7.15) applies, so we need only check that $R^i f_{X/S*}E$ is coherent on S.

Since p is nilpotent on S and the ideal K of S_0 in S is a PD ideal, it is a nil ideal, and since S is noetherian, $K^{m+1} = 0$ for some m. The proof is by induction on m. If $m = 0$, $S_0 = S$, X/S is smooth, and we know from (7.1) that $R^i f_{X/S*}E \cong \mathbb{R}^i f_*(E_X \otimes \Omega^{\cdot}_{X/S})$. Since E_X is coherent, so is each $R^i f_*(E_X \otimes \Omega^p_{X/S})$, and hence so is $R^i f_{X/S*}E$.

Now let $S_n \subseteq S$ be the subscheme defined by K^{n+1}, so that we have an exact sequence:

$$(*) \qquad 0 \to K^n/K^{n+1} \to \mathcal{O}_{S_n} \to \mathcal{O}_{S_{n-1}} \to 0$$

Tensoring with the complex $\mathbb{R}f_{X/S*}(E)$ in the derived category, and using the base change isomorphisms

$$\mathcal{O}_{S_\nu} \overset{\mathbf{L}}{\otimes} \mathbb{R}f_{X/S*}(E) \longleftrightarrow \mathbb{R}f_{X/S_\nu *}(E)$$

for $\nu = n$ and $n-1$, we get a triangle:

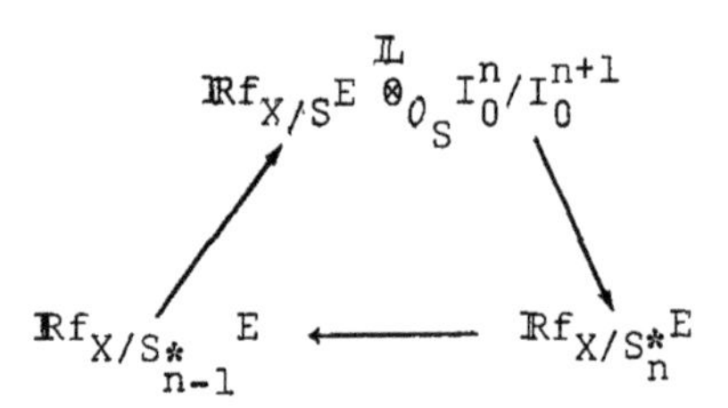

Now the top of the triangle is isomorphic to $\mathbb{R}f_{X/S_0}E\overset{\mathbb{L}}{\otimes}_{\mathcal{O}_{S_0}}I_0^n/I_0^{n+1}$, again by base changing, which has coherent cohomology by the case $m=0$ since $\mathbb{R}f_{X/S_0}{}_*E$ is perfect. By induction, $\mathbb{R}f_{X/S_{n-1}}E$ has coherent cohomology, and the theorem follows from the long exact sequence of the cohomology of a triangle. □

This is all we shall say about crystalline cohomology over a p-adically nilpotent base. What we must do next is construct a p-adic theory by taking an inverse limit. There are several ways to do this; we hope that the reader does not find the one we have chosen too distasteful.

7.17 <u>Definition.</u> A "P-adic base" is a noetherian PD ring (A,I,γ), together with a sub PD ideal P of I such that A is P-adically separated and complete, and such that P contains some prime number p. For each n, we let $A_n = A/P^{n+1}$, $S_n = \mathit{Spec}\ A_n$, $S = \mathit{Spec}\ A$, and $\hat{S} = \mathit{Spf}\ A$ (for the P-adic topology). Furthermore, we let γ denote the induced PD structure on I/P^{n+1}.

Suppose that X/S_0 is a scheme to which γ extends. We want to construct crystalline cohomology of X over $\hat{S}$, compatible with the cohomology of each $(X/S_n)_{cris}$. For each $m \leq n$, we have a morphism: $i_{mn}\colon (X/S_m)_{cris} \to (X/S_n)_{cris}$, compatible with composition. Note that if $T \in \mathrm{Cris}(X/S_n)$, the sheaf $i^*_{mn}(T)$ is representable, in fact by $T \times_{S_n} S_m$ (exercise).

In case X/S_0 is smooth and proper, it is possible to appeal to general properties of limits in the derived category (c.f. (B.11) of Appendix B) to define $\mathbb{R}\Gamma(X/\hat{S}) = \mathrm{R}\varprojlim \mathbb{R}\Gamma(X/S_n)$. This is the approach taken in Berthelot's thesis [CC]. However, in the next chapter, we shall need to work with the crystalline cohomology of nonproper schemes, and therefore we have to give a different construction.

To define the cohomology we want, we construct a site $\mathrm{Cris}(X/\hat{S})$ which computes the limit automatically. Its objects are PD thickenings (U,T,δ) as in §5, such that there exists an n with $p^n \mathcal{O}_T = 0$; its morphisms are defined in the usual way. This is, in fact, the direct limit of the sites $\mathrm{Cris}(X/S_n)$ with the obvious inclusions. For all n, there exists a canonical morphism of topoi $i_n\colon (X/S_n)_{\mathrm{cris}} \to (X/\hat{S})_{\mathrm{cris}}$, for which the inverse image $i^*_n F$ of a sheaf F is just the restriction of F to the subsite $\mathrm{Cris}(X/S_n)$; moreover, for $m \leq n$, $i_n \circ i_{mn} = i_m$.

To deal with the inverse limits arising in computing the cohomology of $(X/\hat{S})_{\mathrm{cris}}$, it is convenient to introduce an auxiliary site $\mathrm{Cris}(X/S.)$ (compare with SGA 4 VI for the general notion of a fibered site). The objects of $\mathrm{Cris}(X/S.)$ are pairs (n,T) such that $n \in \mathbf{N}$ and $T \in \mathrm{Cris}(X/S_n)$; the morphisms $(n',T') \to (n,T)$ are S-PD morphisms $T' \to T$ (and don't exist unless $n' \leq n$), the covering families of (n,T) are just the Zariski coverings $\{(n,T'_i) \to (n,T)\}$. The category

$(X/S.)_{cris}$ of sheaves on $Cris(X/S.)$ can be interpreted as in (5.1): for each $(n,T) \in Cris(X/S.)$, a Zariski sheaf $F_{(n,T)}$ on T, plus "compatibility morphisms"

$$u^{-1}F_{(n,T)} \to F_{(n',T')} \quad \text{if} \quad u\colon (n',T') \to (n,T) .$$

Notice that if $n' < n$, we do not require the map $u^{-1}F_{(n,T)} \to F_{(n',T)}$ to be an isomorphism, in general. We shall say that a sheaf of $\mathcal{O}_{X/S.}$-modules is a "crystal" iff the maps $u^*F_{(n,T)} \to F_{(n',T')}$ are isomorphisms for every u. A sheaf F in $(X/S.)_{cris}$ can also be described as a collection of sheaves $F_n \in (X/S_n)_{cris}$ for all n, together with compatibility maps, as the reader can imagine. Moreover, the functor $F \mapsto F_n$ is the inverse image functor of a morphism of topoi $j_n\colon (X/S_n)_{cris} \to (X/S.)_{cris}$. Finally, there exists a morphism of topoi $j\colon (X/S.)_{cris} \to (X/\hat{S})_{cris}$, such that $(j^*(E))_n = i_n^*E$, i.e. such that $j \circ j_n = i_n$. The formation of the topoi $(X/\hat{S})_{cris}$ and $(X/S.)_{cris}$ are evidently functorial, and we have:

7.18 <u>Proposition.</u> Suppose $(A,I,\gamma) \to (A',I',\gamma')$ is a PD morphism, $P \subseteq I$ is a sub PD ideal, and $PA' \subseteq P'$. Suppose that $f\colon X' \to X$ covers the morphism $S' \to S$. Then f induces, in a natural way, morphisms f_{cris} in the commutative diagram below.

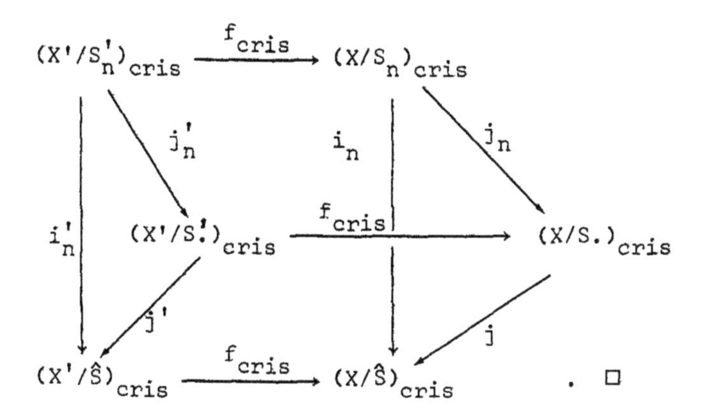

. □

To compare the cohomologies of $(X/\hat{S})_{cris}$ and $(X/S.)_{cris}$, we will use the following result:

7.19 Proposition. If E is an $\mathcal{O}_{X/\hat{S}}$-module in $(X/\hat{S})_{cris}$ and if each $E_{(U,T,\delta)}$ is quasi-coherent, then we have a canonical isomorphism:

$$E \xrightarrow{\cong} \mathbb{R}j_*j^*E \text{ , hence also}$$

$$\mathbb{R}\Gamma(X/\hat{S},E) \xrightarrow{\cong} \mathbb{R}\Gamma(X/S.,j^*E) \text{ .}$$

Proof. First we note the following results concerning j_* and inverse limits.

7.20 Lemma. Let X be a topological space, $F.$ an inverse system of abelian sheaves on X, and suppose that there exists a basis of open sets of X satisfying:

(a) $H^q(V,F_n) = 0$ if $q > 0$

(b) The inverse system $H^0(V,F_n)$ satisfies the Mittag-Leffler condition (ML).

Then $F.$ is $\varprojlim$ acyclic.

<u>Proof.</u> More general statements are well-known, c.f. [EGA 0_{III} 13.3.1] or [LCD, Appendix]. For the readers convenience, we reproduce what we need. A convenient way to view an inverse system indexed by $\mathbb{N}$ is to make the category $\mathbb{N}$ into a site by endowing it with the coarse topology, i.e. only the identity maps are coverings. Then the topos $\tilde{\mathbb{N}}$ of sheaves on $\mathbb{N}$ is just the category of inverse systems, and $\Gamma(\tilde{\mathbb{N}},A) = \varprojlim A_n$. More generally, we can form the site $\mathbb{N} \times X$, whose objects are pairs (n,U), with $U \subseteq X$ open, with the evident morphisms and topology, and a sheaf on $\mathbb{N} \times X$ is the same thing as an inverse system of sheaves on X. Moreover, there exists a morphism of topoi $\pi_X\colon \widetilde{\mathbb{N} \times X} \to \tilde{X}$ such that if $F.$ is such a sheaf, $\pi_{X*}F.$ is just the sheaf $\varprojlim F.$

Suppose now that $F.$ satisfies the hypotheses of the lemma, and that $I.$ is an injective resolution of $F.$ on $\mathbb{N} \times X$. Then $I^q_{\cdot}$ is flasque, so that each I^q_n is flasque, and the maps $I^q_n(U) \to I^q_{n-1}(U)$ are surjective. If V is one of the open sets of the lemma, the complex $\Gamma(V,I^{\cdot}_n)$ is acyclic in positive degrees, and its H^0 is $H^0(V,F_n)$. Thus, the complex, as well as its cohomology, satisfies the Mittag-Leffler condition, hence by [EGA 0_{III} 13.2.3], we find that

$$H^q(\varprojlim \Gamma(V, I^{\cdot}_n)) = \varprojlim H^q(V,I^{\cdot}_n) = 0 \quad \text{if } q > 0,$$

i.e. $\Gamma(V,\pi_{X*}I^{\cdot})$ is acyclic in positive degrees. This proves the lemma. □

7.21 Lemma. If F is an abelian sheaf on $\mathrm{Cris}(X/S.)$:

7.21.1 For $(U,T,\delta) \in \mathrm{Cris}(X/S)$, we have a natural isomorphism:

$$(j_*F)_{(U,T,\delta)} \cong \varprojlim_{n>>0} F_{(n,(U,T,\delta))}$$

7.21.2 If each $F_{(n,(U,T,\delta))}$ is quasi-coherent, and if the maps:

$$F_{(n',(U,T,\delta))} \to F_{(n,(U,T,\delta))} \quad \text{for } n' \geq n$$

are epimorphisms, then F is j_*-acyclic.

Proof. Note that if $(U,T,\delta) \in \mathrm{Cris}(X/\hat{S})$, $(n,(U,T,\delta)) \in \mathrm{Cris}(X/S.)$ for $n >> 0$, so that (7.21.1) makes sense. If $F \in \mathrm{Cris}(X/S.)$, we get an inverse system $F._{(U,T,\delta)}$ of sheaves on T by

$$F_{n,(U,T,\delta)} = F_{(n,(U,T,\delta))} = (j_n^*F)_{(U,T,\delta)} \quad (\text{for } n >> 0).$$

To verify that $\varprojlim_n F_{n,(U,T,\delta)} = j_*F$, one has only to check the universal mapping property, which is immediate.

Now fix $T = (U,T,\delta)$, and observe that we have a commutative diagram:

$$\begin{array}{ccc} (X/S.)_{cris} & \longrightarrow & \widetilde{\mathbb{N}} \times T_{zar} \\ \downarrow j_* & & \downarrow \pi_{T_*} \\ (X/\hat{S})_{cris} & \longrightarrow & T_{zar} \end{array} \qquad \begin{array}{l} F \mapsto \{F_{n,(U,T,\delta)}: n >> 0\} \\ \\ G \longmapsto G_{(U,T,\delta)} \end{array}$$

By the analogue of (5.26.3), the two horizontal arrows are exact and take injectives to injectives. Thus, if $F \in (X/S.)_{cris}$ is abelian, $(R^q j_* F)_T = (R^q \pi_{T*})(F._T)$, so to prove (7.21.2), it suffices to see that $F._T$ is acyclic for π_{T*}. This follows immediately from (7.20). □

To prove Proposition (7.19), note simply that if $E \in (X/\hat{S})_{cris}$, then $(j^*E)_{(n,(U,T,\delta))} \cong E_{(U,T,\delta)}$ for all $n >> 0$, so (7.21.2) is certainly satisfied, and $(j_*j^*E)_{(U,T,\delta)} = \varprojlim_n (j^*E)_{(n,(U,T,\delta))} = E_{(U,T,\delta)}$. □

Because of their functoriality, the constructions of $(X/\hat{S})_{cris}$ and of $(X/S.)_{cris}$ are compatible with localization in the Zariski topology of X, and we can define morphisms $u_{X/\hat{S}}: (X/\hat{S})_{cris} \to X_{zar}$ and $u_{X/S.}: (X/S.)_{cris} \to X_{zar}$ exactly as in (5.18). There is even a morphism $u_{X/S.}: (X/S.)_{cris} \to \widetilde{\mathbb{N}} \times X_{zar}$, with $u_{X/S.*}(F)(n,U) = \mathbb{R}\Gamma(U/S_n, j_n^* F)$, and a commutative diagram:

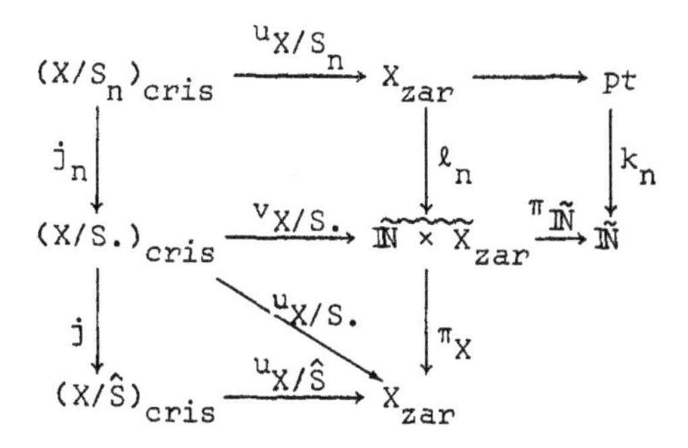

7.22 <u>Proposition.</u> Let $\gamma = \pi_{\widetilde{\mathbb{N}}} \circ v_{X/S.}$ in the above diagram.

7.22.1 If F is an abelian sheaf in $(X/S.)_{cris}$,

$$(\mathbb{R}\gamma_* F)_n \cong \mathbb{R}\Gamma(X/S_n, j_n^* F), \quad \text{and}$$

$$\ell_n^* (\mathbb{R}v_{X/S.*} F) \cong \mathbb{R}u_{X/S_n *}(j_n^* F) .$$

7.22.2 If E is a quasi-coherent sheaf in $(X/\hat{S})_{cris}$, then

$$\mathbb{R}\Gamma(X/\hat{S}, E) \cong \mathbb{R}\varprojlim \mathbb{R}\Gamma(X/S_n, i_n^* E),$$

and

$$\mathbb{R}u_{X/\hat{S}*} E \cong \mathbb{R}\varprojlim \mathbb{R}u_{X/S_n *}(i_n^* E).$$

<u>Proof.</u> We had best begin by remarking that we have abused language in the second statement, since we have written $\mathbb{R}\varprojlim$ in a derived categroy. However, the diagram and (7.22.1) gives it meaning, for $\mathbb{R}\gamma_* j^* E$ will be a well-defined object in the derived category of abelian sheaves on $\mathbb{N}$, and $(\mathbb{R}\gamma_* j^* E)_n = \mathbb{R}\Gamma(X/S_n, i_n^* E)$ for all n. Moreover, by (7.19), $\mathbb{R}\Gamma(X/\hat{S}, E) \cong \mathbb{R}\Gamma(X/\hat{S}, \mathbb{R}j_* j^* E) \cong \mathbb{R}\Gamma(X/S., j^* E)$, so (7.22.2) will follow from (7.22.1).

We first observe that the functor $j_n^*: (X/S.)_{cris} \to (X/S_n)_{cris}$ has an exact left adjoint, given (for abelian sheaves, of course) by $[j_n!(G)]_m = \{0$ if $m > n$, $j_m^*(G)$ if $m \leq n\}$. The same is true of ℓ_n^* and k_n^*, and it follows that they take injectives to injectives. Since they are also exact, they satisfy $\mathbb{R}j_n^* = j_n^*$, etc. We conclude that $\ell_n^* \mathbb{R}v_{X/S.*}F \cong \mathbb{R}\ell_n^* \mathbb{R}v_{X/S.*}F$ $\cong \mathbb{R}(\ell_n^* \circ v_{X/S.*})F \cong \mathbb{R}(u_{X/S_n*} j_n^*) \cong \mathbb{R}u_{X/S.*} \circ \mathbb{R}j_n^* F \cong \mathbb{R}u_{X/S_n*}(j_n^* F)$. The rest of the proposition is proved similarly. □

We are especially interested in the cohomology of a crystal of $\mathcal{O}_{X/\hat{S}}$-modules. If X can be embedded in a smooth S-scheme Y, we can obtain an "explicit" complex representing this cohomology.

7.23 <u>Theorem.</u> Suppose Y/S is smooth, $X \hookrightarrow Y$ is closed, and E is a quasi-coherent crystal of $\mathcal{O}_{X/\hat{S}}$-modules. Let $D = D_{X,Y}(Y)$, and let $\hat{D}$ be its P-adic completion. Then there is a $\hat{D}$-module with integrable connection (E,∇) inducing E, and for each m, there is a natural isomorphism:

$$\mathbb{R}u_{X/\hat{S}*} J_{X/\hat{S}}^{[m]} E \to F_X^m(E \otimes \Omega^\cdot_{\hat{D}/\hat{S}})$$

<u>Proof.</u> Since $u_{X/\hat{S}} \circ j = u_{X/S.}$, we have a canonical isomorphism: $\mathbb{R}u_{X/\hat{S}*} \mathbb{R}j_* j^*(J_{X/\hat{S}}^{[m]} E) \xrightarrow{\cong} \mathbb{R}u_{X/S.*}(j^* J_{X/\hat{S}}^{[m]} E)$. But (7.19) tells us that $\mathbb{R}j_* j^*(J_{X/\hat{S}}^{[m]} E) \cong J_{X/\hat{S}}^{[m]} E$, and one sees easily that $j^*(J_{X/\hat{S}}^{[m]} E) \cong J_{X/S.}^{[m]}(j^*E)$, and that j^*E is a crystal on Cris(X/S.). Hence it suffices to prove the theorem with (X/S.) in place of $(X/\hat{S})$.

For each n, the crystal E_n on $\mathrm{Cris}(X/S_n)$ corresponds to a $\mathcal{D}_n = \mathcal{D}_{X,Y}(Y\times_S S_n)$-module with integrable connection $(\mathcal{E}_n,\nabla)$, with $\mathcal{E}_n = E_{D_n}$, by (6.6). Since $\mathcal{D}_n \otimes_A A_{n-1} = \mathcal{D}_{n-1}$ by (3.20.8), $\mathcal{E}_n \otimes_A A_{n-1} = \mathcal{E}_{n-1}$ also. Let $\mathcal{D}.$ denote the sheaf on $\mathbb{N}\times X$ corresponding to $\{\mathcal{D}_n\}$, and let $(\mathcal{E}.,\nabla)$ the sheaf of $\mathcal{D}.$-modules with connection corresponding to $\{\mathcal{E}_n\}$. Then $\hat{\mathcal{D}} = \pi_{X*}\mathcal{D}.$, and $(\mathcal{E},\nabla) = (\pi_{X*}\mathcal{E}.,\nabla)$ is a $\hat{\mathcal{D}}$-module with integrable connection, inducing E.

Even though we have compatible isomorphisms $\mathbb{R}u_{X/S_n*}J^{[m]}_{X/S_n} E_n \to F^m(\mathcal{E}_n \otimes \Omega^{\bullet}_{D_n/S_n})$ for each n, this doesn't quite give us the theorem, because one cannot work locally in the derived category. Therefore we must copy over the proof of (7.2).

Let $D.$ denote the sheaf in $(X/S.)_{\mathrm{cris}}$ defined by $\{D_n : n \in \mathbb{N}\}$, and form the diagram below (as in (6.10)).

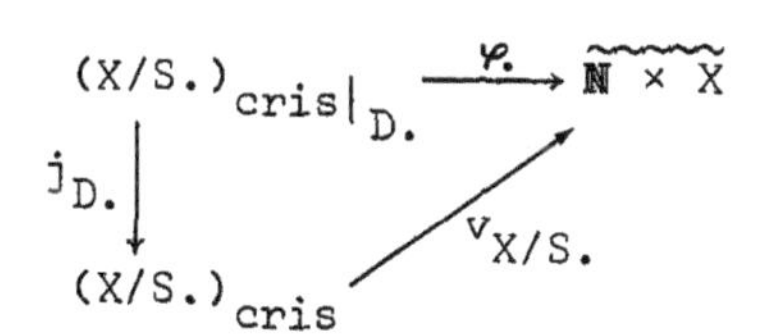

I claim that:

7.23.1 $\varphi._*$ and $j_{D.*}$ are exact.

7.23.2 If F is any abelian sheaf in $(X/S.)_{\mathrm{cris}|D.}$, $j_{D.*}F$ is acyclic for $v_{X/S.*}$.

Indeed, it suffices to check (7.23.1) after restricting to (X/S_n) for all n, so that it follows from (5.26) and (5.27). Moreover, (7.23.1) implies (7.23.2) (c.f. (5.27.2)).

There is a natural morphism of complexes: $J^{[m]}_{X/S.}E \to F^m_X L(E.\otimes\Omega^{\cdot}_{D./S})$ where the latter in degree q is $K^{[m-q]}_{\cdot}L(E.\otimes\Omega^q_{D./S}) \cong j_{D.*}(J^{[m-q]}_{X/S}\varphi^*_{\cdot}E.\otimes\Omega^q_{D./S})$ (c.f. 6.13). It is a quasi-isomorphism because it is after restricting to each $(X/S_n)_{cris}$, by (6.13), and hence we obtain an isomorphism:

$$\mathbb{R}v_{X/S.*}J^{[m]}_{X/S.}E \xrightarrow{\cong} v_{X/S.*}F^m_X L(E.\otimes\Omega^q_{D./S}) \cong F^m_X(E.\otimes\Omega^{\cdot}_{D./S}) .$$

Since (7.21) tells us that the sheaves of the complex $F^m_X(E.\otimes\Omega^{\cdot}_{D./S})$ on $\mathbb{N} \times X$ are acyclic for $\mathbb{R}\pi_{X*}$, the theorem follows. □

We now come to the main statement of P-adic crystalline cohomology.

7.24 <u>Theorem.</u> Suppose X/S_0 is smooth, quasi-compact, and quasi-separated, and E is a locally free, finitely generated crystal of $\mathcal{O}_{X/\hat{S}}$-modules in $(X/\hat{S})_{cris}$. Then:

7.24.1 There is an object D in the derived category of inverse systems of A-modules such that each $D_n \cong \mathbb{R}\Gamma(X/S_n, i^*_n E)$ and $\mathbb{R}\varprojlim D \cong \mathbb{R}\Gamma(X/\hat{S},E)$.

7.24.2 For each n, there is a natural base-changing isomorphism:

$$\mathbb{R}\Gamma(X/S,E) \overset{\mathbb{L}}{\underset{A}{\otimes}} A_n \longrightarrow \mathbb{R}\Gamma(X/S_n, i_n^* E) .$$

Moreover, $\mathbb{R}\Gamma(X/S,E)$ has finite tor-dimension.

7.24.3 If X/S_0 is proper, the complex $\mathbb{R}\Gamma(X/S,E)$ is perfect. Moreover, the inverse systems $H^i(X/S_n, i_n^* E)$ satisfy ML, and there is an isomorphism:

$$H^i(X/\hat{S},E) \to \varprojlim H^i(X/S_n, i_n^* E) .$$

<u>Proof.</u> The first statement is just (7.22.2) (and in fact is true more generally). Now the base changing theorem (7.8) implies that the arrows $D_n \overset{\mathbb{L}}{\underset{A_n}{\otimes}} A_{n-1} \to D_{n-1}$ are (quasi-) isomorphisms for all n, so that D is what Appendix B calls a "quasi-consistent complex of $A.$-modules" (c.f. B4). Then B6) and (7.13) imply that D and $\mathbb{R}\varprojlim D$ have finite tor-dimension, and (B5) implies that the arrows $(\mathbb{R}\varprojlim D) \overset{\mathbb{L}}{\underset{A}{\otimes}} A_n \to D_n$ are quasi-isomorphisms. This proves (7.24.2). If X_0/S_0 is proper, we know that $\mathbb{R}\Gamma(X_0/S_0, E_0)$ is perfect, and so (B11) implies that $\mathbb{R}\Gamma(X/\hat{S}, E)$ is also perfect. Let us remark that (B11) shows that in the proper case only, $\mathbb{R}\Gamma(X/\hat{S},E)$ is functorially determined by the inverse system $\mathbb{R}\Gamma(X/S_n, i_n^* E)$. □

7.25 <u>Corollary.</u> Suppose that in the notation of (7.17) above, A is a complete discrete valuation ring (necessarily of mixed characteristic p and absolute ramification index $\leq p-1$) with

$P = I =$ the maximal ideal of A. If E and X are as in (7.25), there are exact sequences:

$$0 \to H^i(X/\hat{S}, E) \otimes A_n \to H^i(X/S_n, E_n) \to \mathrm{Tor}_1^A(H^{i+1}(X/\hat{S}, E), A_n) \to 0 .$$

Proof. Let $K^{\cdot}$ be a strictly perfect complex A-modules representing $\mathbb{R}\Gamma(X/S.,E)$. Let π be the uniformizing parameter of A. There is an exact sequence:

$$0 \to K^{\cdot} \xrightarrow{\pi^{n+1}} K^{\cdot} \to K^{\cdot} \otimes A_n \to 0 .$$

Hence:

$$H^i(K^{\cdot}) \xrightarrow{\pi^{n+1}} H^i(K^{\cdot}) \to H^i(K^{\cdot} \otimes A_n) \to H^{i+1}(K^{\cdot}) \xrightarrow{\pi^{n+1}} H^{i+1}(K^{\cdot}) .$$

Since $H^*(K^{\cdot}) \cong H^*(X/\hat{S}, E)$ and $H^i(K^{\cdot} \otimes A_n) \cong H^i(X/S_n, E_n)$, the result is clear. □

We have at last succeeded in giving at least the definition of a reasonable p-adic Weil cohomology. Here is what we have proved:

7.26 Summary. Suppose A is as in (7.25). Then there is a functor H^*_{cris} from the category of smooth proper A_0-schemes to the category of finitely generated graded A-modules such that

7.26.1 $$H^*_{cris}(X) \underset{def}{=} \mathbb{H}^*(X/\hat{S}, \mathcal{O}_{X/\hat{S}}) .$$

7.26.2 There are natural exact sequences:

$$0 \to H^i_{cris}(X) \otimes_A A_0 \to H^i_{DR}(X/S_0) \to \mathrm{Tor}_1^A(H^{i+1}_{cris}(X), A_0) \to 0 .$$

7.26.3 If Y/S is a smooth lifting of X, there is a natural isomorphism:

$$H^i_{cris}(X) \cong H^i_{DR}(Y/S) .$$

<u>Proof.</u> It remains only to explain the last statement. But $H^i_{cris}(X) = \varprojlim H^i(X/S_{n\,cris}, \mathcal{O}_{X/S_n}) \cong \varprojlim H^i(Y/S_{n\,cris}, \mathcal{O}_{Y/S_n})$, by (7.4), which is turn is the same as $\mathbb{H}^i(Y/S, \Omega^{\bullet}_{Y/S})$ by the fundamental theorem for a proper morphism. □

In particular, we obtain the desired relationship between the crystalline and DeRham Betti numbers. Explicitly, if β_i is the rank of the free part of $H^i_{cris}(X/S.)$ and if η_i is the number of its torsion factors, then the dimension of $H^i_{DR}(X_0/k)$ is $\beta_i + \eta_i + \eta_{i+1}$. □

References for Chapter 7

[ADRC] Hartshorne, R. "On the De Rham Cohomology of Algebraic Varieties" Pub. Math. I.H.E.S. No. 45 (1976) 5-99.

[CC] Berthelot, P. "Cohomologie cristalline des schémas de caractéristique p > 0" Lecture Notes in Mathematics No. 407, Springer Verlag (1976).

[EGA III] Grothendieck, A and Dieudonné, J. "Eléments de Géométrie Algébrique" Publ. Math. I.H.E.S. No. 11 (1961).

[H III] Deligne, P. Théorie de Hodge III Publ. Math. I.H.E.S. No. 44 (1975) 5-77.

[LCD] Ogus, A. "Local Cohomological Dimension of Algebraic Varieties" Ann. of Math. 98 (1973) 327-365.

[SGA 4;6] Grothendieck, A et. al. "Séminaire de Géométrie Algébrique" Lecture Notes in Mathematics No.'s 269, 270, 305; 225, Springer Verlag.

§8. Frobenius and the Hodge Filtration.

Suppose W is the Witt ring of a perfect field k of characteristic p, and X is a smooth k-scheme. The Frobenius automorphism of W is a PD morphism, covered by the absolute Frobenius endomorphism F_X of X, and it follows that F_X acts on the crystalline cohomology of X relative to W. In this chapter we shall study this action, in particular, its relationship to the Hodge filtration on crystalline cohomology (as determined from the ideal $J_{X/S}$). The main global applications are Mazur's theorem (8.26), which says that (with suitable hypotheses on X) the action of Frobenius determines the Hodge filtration on $H^*_{DR}(X/k)$, and Katz's conjecture (8.39), which says how the Hodge filtration limits the possible "slopes" of Frobenius.

The above results generalize somewhat the work of Mazur [4,5]. Our technique of proof is, however, rather different, since we follow a suggestion of Deligne, proving a local result of which the above global statements are formal consequences.

We approach the local problem in two parts. In the first part we study the DeRham cohomology of a lifting to obtain a statement (18.3) and its generalization (8.8) only valid in the liftable case. In the second part we interpret the calculations in terms of crystalline cohomology and obtain (8.20), which does have global meaning.

(8.1) Let us fix some notation. Let (A, I, γ) be a P-adic base (7.17), and use the notations of (7.17) except write $S = \mathcal{S}pf\ A$ instead of $S = \mathcal{S}pec\ A$. We assume as additional hypotheses:

$P = (p)$, and A is a p-torsion free. In addition, we give ourselves an endomorphism F_S of S, lifting the absolute Frobenius endomorphism F_{S_0} of S_0.

(8.2) The local calculations take place in what we shall call a "lifted situation over (S, F_S)". This means a formally smooth formal scheme Y/S, together with an F_S-morphism $F_Y\colon Y \to Y$ such that $F_Y \times_S S_0 = F_{Y_0}$, the absolute Frobenius endomorphism of $Y_0 = Y \times_S S_0$.

We shall also need "relative Frobenius", whose construction we recall. Form the fiber product $Y' = Y \times_S S$, using the map $F_S\colon S \to S$. Let $pr\colon Y \to S$, $pr'\colon Y' \to S$, and $W_{Y/S}\colon Y' \to Y$ be the natural projections, and let $F_{Y/S}\colon Y \to Y'$ be the S-map such that $W_{Y/S} \circ F_{Y/S} = F_Y$. Thus we have a diagram:

$$
\begin{array}{ccccccc}
Y & \xrightarrow{F_{Y/S}} & Y' & \xrightarrow{W_{Y/S}} & Y & \xrightarrow{F_{Y/S}} & Y' \\
& {\scriptstyle pr}\searrow & \downarrow{\scriptstyle pr'} & & \downarrow{\scriptstyle pr} & \swarrow{\scriptstyle pr'} & \\
& & S & \xrightarrow{F_S} & S & &
\end{array}
\tag{8.2.1}
$$

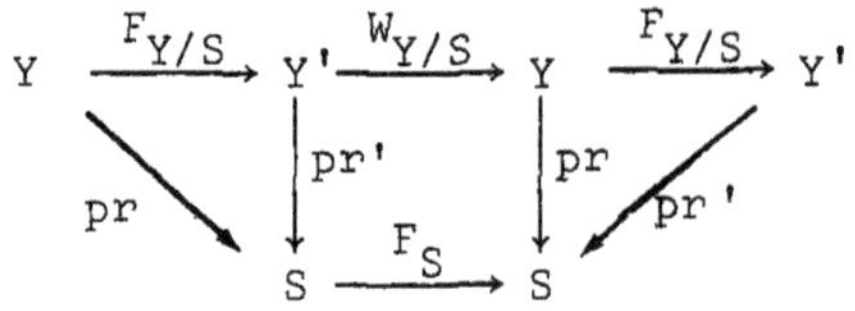

Notice that $F_{Y/S}$ and $W_{Y/S}$ are homeomorphisms. Indeed, since we are working with p-adic formal schemes, it is enough to check this mod p, and we have $W_{Y_0/S_0} \circ F_{Y_0/S_0} = F_{Y_0}$ and $F_{Y_0/S_0} \circ W_{Y_0/S_0} = F_{Y'_0}$. Notice also that the above diagram makes sense for any Y/S, not necessarily smooth.

Let me now explain a special case of the main local result, which gives a precise description of the image of $F^*_{Y/S}\colon \Omega^\bullet_{Y'/S} \to F_{Y/S*}\Omega^\bullet_{Y/S}$. Begin by noting that $F^*_{Y_0/S_0}$ kills $\Omega^1_{Y'_0/S_0}$, so that the image of $\Omega^1_{Y'/S}$ is contained in

$pF_{Y/S_*}\Omega^1_{Y/S} = F_{Y/S_*}p\Omega^1_{Y/S}$. Since $F^*_{Y/S}$ is compatible with exterior products, it follows that the image of $\Omega^k_{Y'/S}$ is contained in $p^kF_{Y/S*}\Omega^k_{Y/S}$. We can say slightly more than this: Since $F^*_{Y/S}$ is a morphism of complexes, it follows that the image of $\Omega^k_{Y'/S}$ is in fact contained in $\{\omega \in p^kF_{Y/S_*}\Omega^k_{Y/S}: \; d\omega \in p^{k+1}F_{Y/S_*}\Omega^{k+1}_{Y/S}\}$. Our main result says, in a sense, that this is exactly the image of F_{Y/S_*} . Precisely:

8.3 Theorem. In a lifted situation $(Y/S, F_Y)$, the map $F^*_{Y/S}: \; \Omega^\cdot_{Y'/S} \to F_{Y/S_*}\Omega^\cdot_{Y/S}$ induces a quasi-isomorphism into the largest subcomplex $N^\cdot$ of $F_{Y/S_*}\Omega^\cdot_{Y/S}$ such that $N^k \subseteq p^kF_{Y/S_*}\Omega^k_{Y/S}$ for all k (described explicitly above).

Both for the proof and applications, we shall in fact need to make a more general statement, (8.8). Let us first explain the main tools of the proof, which are the Cartier isomorphism and its relationship to Frobenius. For the reader's convenience, we recall the following description of the Cartier isomorphism, proved in [2, 7.2].

8.4 Theorem (Cartier). Suppose X/S_0 is smooth. Then there is a unique morphism of $\mathcal{O}_{X'}$-modules:

$$C^{-1}: \Omega^i_{X'/S_0} \to H^i(F_{X/S_0*}\Omega^\cdot_{X/S_0})$$

such that:

$C^{-1}(1) = 1$, $C^{-1}(\omega\wedge\tau) = C^{-1}(\omega)\wedge C^{-1}(\tau)$, and $C^{-1}(dW^*_{X/S_0}(\alpha)) =$ the image in $H^1(F_{X/S_0}*\Omega^\cdot_{X/S_0})$ of $\alpha^{p-1}d\alpha$, for any section α of $\mathcal{O}_{X/S_0}$.

Furthermore, C^{-1} is an isomorphism. □

Here is the relationship between the action of Frobenius and C^{-1} in a lifted situation.

8.5 Theorem. Suppose $(Y/S, F_Y)$ is a lifted situation (8.2). Then for each $j \geq 0$, there is a commutative diagram

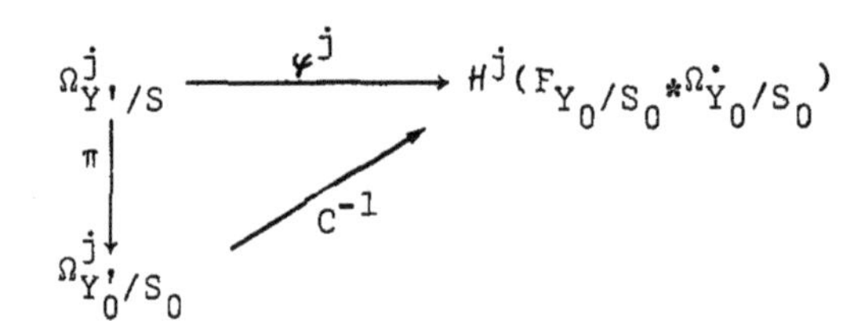

where π is "reduction mod p" and $\varphi^j(\omega)$ is the image in $\mathcal{H}^j$ of $p^{-j}F^*_{Y/S}(\omega)$.

Proof. We have already observed that $F^*_{Y/S}(\omega)$ lies in $\{\theta \in p^j F_{Y/S_*}\Omega^j_{Y/S}\colon\ d\theta \in p^{j+1}F_{Y/S_*}\Omega^{j+1}_{Y/S}\}$; it follows that $p^{-j}F^*_{Y/S}(\omega)$ makes sense and that its image in $\Omega^j_{Y_0/S_0}$ is a cocycle. To see that the diagram commutes, first observe that $C^{-1}\circ\pi$ and φ^j are both $\mathcal{O}_{Y'}$-linear. Moreover, if $j=0$ both map 1 to 1, so the diagram commutes. If $j \geq 1$, $\Omega^j_{Y'/S} \cong W^*_{Y/S}(\Omega^j_{Y/S})$ is generated as an $\mathcal{O}_{Y'}$-module by elements of the form $\omega = \omega_1 \wedge \ldots \wedge \omega_j$, with $\omega_i \in \Omega^1_{Y'/S}$. We can even take $\omega_i = dW^*_{Y/S}(\alpha_i)$, with α_i a section of $\mathcal{O}_Y$. Then $\varphi^j_{Y/S}(\omega)$ is the class of

$$p^{-j}F^*_{Y/S}(dW^*_{Y/S}(\alpha_1) \wedge \ldots dW^*_{Y/S}(\alpha_j)) = p^{-j}\, dF^*_Y(\alpha_1) \wedge \ldots \wedge dF^*_Y(\alpha_j) .$$

Now $F^*_Y(\alpha_i) = \alpha_i^p + p\beta_i$ for some β_i, so $dF^*_Y(\alpha_i) = p\alpha_i^{p-1}d\alpha_i + pd\beta_i$. Thus, $\varphi^j(\omega)$ is the class of $(\alpha_1^{p-1}d\alpha_1 + d\beta_1)\wedge\ldots\wedge(\alpha_j^{p-1}d\alpha_j + d\beta_j)$, which is the same as the class of $(\alpha_1^{p-1}d\alpha_1)\wedge\dots\wedge(\alpha_j^{p-1}d\alpha_j)$. By (8.4), this is $C^{-1}(\omega_1)\wedge\dots\wedge C^{-1}(\omega_j)$. □

In order to state the generalization of Theorem (8.8) which we will need, we must first introduce some notation which permits us to describe the p-adic divisibility of a morphism of complexes. For the moment, the following ad hoc definition will do, later we shall use a more systematic notion (8.15 ff).

8.6 Definition. Let p be a fixed prime number, and let A be an abelian category. If $\varepsilon: \mathbb{Z} \to \mathbb{N}$ is a function and if $K^\bullet$ is a complex in A, then "$K_\varepsilon^\bullet$" denotes the subcomplex of $K^\bullet$ given by:

$$K_\varepsilon^i = p^{\varepsilon(i)}K^i \cap d^{-1}[p^{\varepsilon(i+1)}K^{i+1}] \quad .$$

Thus, $K_\varepsilon^\bullet$ is the largest subcomplex of $K^\bullet$ such that $K_\varepsilon^i \subseteq p^{\varepsilon(i)}K^i$ for all i.

Notice that if $K^\bullet$ is a complex of p-torsion free sheaves of abelian groups on some site, then $K_\varepsilon^i(U) = \{x \in p^{\varepsilon(i)}K^i(U) : dx \in p^{\varepsilon(i+1)}K^{i+1}(U)\}$, for each object U. (In the presence of torsion, this may only define a presheaf, and K_ε^i is the associated sheaf.)

It is clear that a morphism $f: A^\bullet \to B^\bullet$ induces a morphism $f_\varepsilon: A_\varepsilon^\bullet \to B_\varepsilon^\bullet$ for all ε, and that this defines a functor. It is equally clear that if $\varepsilon \leq \varepsilon'$ (i.e. if $\varepsilon(i) \leq \varepsilon'(i)$ for all i), there is a natural inclusion $A_{\varepsilon'}^\bullet \to A_\varepsilon^\bullet$. This (together with the evident compatibilities) is all we shall need for the present.

8.7 Definition. A function $\varepsilon: \mathbb{Z} \to \mathbb{N}$ is called a "gauge" iff $\varepsilon(i) - 1 \leq \varepsilon(i+1) \leq \varepsilon(i)$ for all i. (This is slightly different from Mazur's notion in [5].)

We can now state and prove the first formulation of the main local result. It will turn out only to have crystalline meaning for what are called "tame gauges", which we shall discuss later.

8.8 Theorem. Let $(Y/S, F_Y)$ be a lifted situation (8.2), and let $\Psi_{Y/S}: \Omega^{\cdot}_{Y'/S} \to F_{Y/S_*}\Omega^{\cdot}_{Y/S}$ be the morphism induced by relative Frobenius $F_{Y/S}$. If $\varepsilon: \mathbb{Z} \to \mathbb{N}$ is any gauge, set $\eta(i) = \varepsilon(i) + i$ for all i. Then there is a natural commutative diagram, in which Ψ_ε is a quasi-isomorphism:

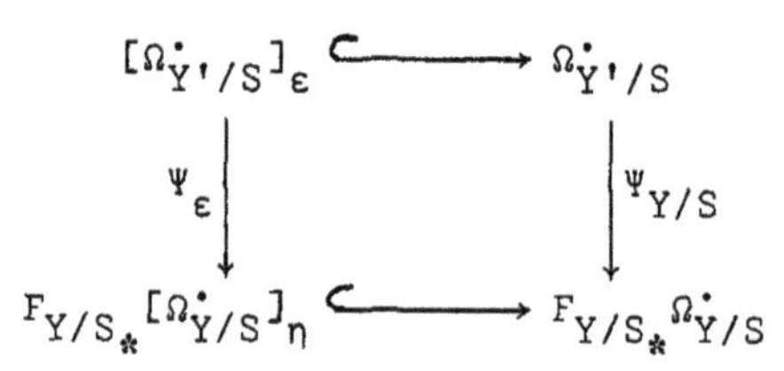

Proof. The existence of this diagram is clear. We shall prove that Ψ_ε is a quasi-isomorphism by unscrewing the gauge ε, always relying on the Cartier isomorphism (8.4) and its relation to $\Psi_{Y/S}$.

8.9 Definition. Suppose ε and ε' are gauges and j is an integer. We say that ε' is a "simple augmentation of ε at j" iff $\varepsilon'(i) = \varepsilon(i)$ whenever $i \neq j$, but $\varepsilon'(j) = \varepsilon(j) + 1$. We have a picture:

8.10 <u>Lemma.</u> Suppose ε' is a simple augmentation of ε at j. Then there is a commutative diagram:

$$\begin{array}{ccc} [\Omega^{\cdot}_{Y'/S}]_{\varepsilon}/[\Omega^{\cdot}_{Y'/S}]_{\varepsilon'} & \xrightarrow{\psi_{\varepsilon}} & F_{Y/S*}[\Omega^{\cdot}_{Y/S}]_{\eta}/[\Omega^{\cdot}_{Y/S}]_{\eta'} \\ \pi_j \downarrow \cong & & \downarrow \rho_j \\ \Omega^{j}_{Y_0'/S_0}[-j] & \xrightarrow{C^{-1}} & H^j(F_{Y_0/S_0*}\Omega^{\cdot}_{Y/S_0})[-j] \end{array}$$

in which π_j is $p^{-\varepsilon(j)}$ followed by reduction mod p, ψ_ε is induced by Ψ_ε, and ρ_j is a quasi-isomorphism.

<u>Proof.</u> Since ε is decreasing, $[\Omega^i_{Y'/S}]_\varepsilon = p^{\varepsilon(i)}\Omega^i_{Y'/S}$ (and similarly for ε'). Thus, $[\Omega^i_{Y'/S}]_{\varepsilon'} = [\Omega^i_{Y'/S}]_\varepsilon$ if $i \neq j$, and we get the isomorphism π_j.

Now set $n = j + \varepsilon(j) = \eta(j) = \eta(j-1)$, and notice that $[\Omega^i_{Y/S}]_{\eta'} = [\Omega^i_{Y/S}]_\eta$ for $i \neq j-1, j$. I claim that there is a commutative diagram:

$$\begin{array}{ccccc} [\Omega^{j-1}_{Y/S}]_{\eta}/[\Omega^{j-1}_{Y/S}]_{\eta'} & \xrightarrow{d} & [\Omega^{j}_{Y/S}]_{\eta}/[\Omega^{j}_{Y/S}]_{\eta'} & & \\ \cong \downarrow \text{"} d \circ p^{-n} \text{"} & & \cong \downarrow p^{-n} & & \\ B^j_{Y_0/S_0} & \hookrightarrow & Z^j_{Y_0/S_0} & \longrightarrow & H^j(\Omega^{\cdot}_{Y_0/S_0}) \end{array}$$

which evidently proves that the arrow ρ_j is a quasi-isomorphism. From the definitions:

$$[\Omega^{j-1}_{Y/S}]_\eta/[\Omega^{j-1}_{Y/S}]_{\eta'} = \{\omega \in p^n\Omega^{j-1}_{Y/S}\}/\{\omega \in p^n\Omega^{j-1}_{Y/S} : d\omega \in p^{n+1}\Omega^j_{Y/S}\} ,$$

so that the arrow induced by $d \circ p^{-n}$ is an isomorphism. Similarly,

$$[\Omega^{j}_{Y/S}]_\eta/[\Omega^{j}_{Y/S}]_{\eta'} = \{\omega \in p^n\Omega^{j}_{Y/S} : d\omega \in p^{n+1}\Omega^{j+1}_{Y/S}\}/\{\omega \in p^{n+1}\Omega^j_{Y/S}\} ,$$

so that the vertical arrow induced by p^{-n} in the diagram is a quasi-isomorphism.

To complete the proof of the lemma, we have only to check that the diagram commutes, i.e. that $C^{-1}\pi_j = \rho_j\psi_\varepsilon$ – which follows immediately from (8.5). □

8.11 Lemma. Suppose ε and ε' are two gauges such that $\varepsilon \le \varepsilon'$ and $\varepsilon(i) = \varepsilon'(i)$ for almost all i. Then Ψ_ε is an isomorphism iff $\Psi_{\varepsilon'}$ is.

Proof. First suppose that ε' is a simple augmentation of ε at j. Then Lemma (8.10) gives us a commutative diagram:

$$\begin{array}{ccccccccc}
0 \longrightarrow & [\Omega^{\cdot}_{Y'/S}]_{\varepsilon'} & \longrightarrow & [\Omega^{\cdot}_{Y'/S}]_{\varepsilon} & \xrightarrow{\pi_j} & \Omega^{j}_{Y_0'/S_0}[-j] & \longrightarrow 0 \\
& \Big\downarrow{\scriptstyle \Psi_{\varepsilon'}} & & \Big\downarrow{\scriptstyle \Psi_{\varepsilon}} & & \Big\downarrow{\scriptstyle C^{-1}} & \\
0 \longrightarrow & F_{Y/S_*}[\Omega^{\cdot}_{Y/S}]_{\eta'} & \longrightarrow & F_{Y/S_*}[\Omega^{\cdot}_{Y/S}]_{\eta} & \xrightarrow{\rho_j} & H^j(F_{Y_0/S_0*}\Omega^{\cdot}_{Y_0/S_0})[-j] & \longrightarrow 0
\end{array}$$

Since C^{-1} is an isomorphism, we see from the long exact sequence of cohomology and the five-lemma that Ψ_ε is a quasi-isomorphism iff $\Psi_{\varepsilon'}$ is.

It follows by induction that if we can find a chain of simple augmentations $\varepsilon = \varepsilon_0 \le \varepsilon_1 \le \cdots \le \varepsilon_n = \varepsilon'$, then the lemma is true for ε and ε'. Thus, the following lemma completes the proof.

8.12 Lemma. Suppose that $\varepsilon \le \varepsilon'$ and $\varepsilon(i) = \varepsilon'(i)$ for almost all i. Then there is a chain of simple augmentations:
$\varepsilon = \varepsilon_0 < \varepsilon_1 < \cdots < \varepsilon_n = \varepsilon'$.

Proof. Since there are only a finite number of δ with $\varepsilon \leq \delta \leq \varepsilon'$, it is enough to prove that there exists a simple augmentation ε_1 of ε such that $\varepsilon < \varepsilon_1 \leq \varepsilon'$ (provided, of course, $\varepsilon \neq \varepsilon'$). It is clear that we can find a simple augmentation of ε at j which is still a gauge iff $\varepsilon(j-1) > \varepsilon(j) = \varepsilon(j+1)$, like this:

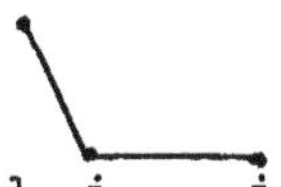

j-1 j   j+1

Let us call such a j a "dent" of ε. We have to show that there is a dent j of ε with $\varepsilon(j) < \varepsilon'(j)$.

Suppose this is not the case. Since $\varepsilon \neq \varepsilon'$, there exists a k such that $\varepsilon(k) < \varepsilon'(k)$. By assumption, k is not a dent of ε, and this can only be for one of two reasons:

(a) $\varepsilon(k+1) < \varepsilon(k)$. But then $\varepsilon(k+1) = \varepsilon(k)-1 < \varepsilon'(k)-1 \leq \varepsilon'(k+1)$, and so also $k' = k+1$ is not a dent of ε. Since $\varepsilon(k'-1) > \varepsilon(k')$, this can only be because $\varepsilon(k'+1) < \varepsilon(k')$. Now we can repeat the argument, and we obtain the double absurdity: ε is strictly decreasing beyond k and $\varepsilon \neq \varepsilon'$ beyond k.

(b) $\varepsilon(k-1) = \varepsilon(k)$. Then $\varepsilon(k-1) = \varepsilon(k) < \varepsilon'(k) \leq \varepsilon'(k-1)$, so $k' = k-1$ is not a dent of ε. Since $\varepsilon(k'+1) = \varepsilon(k')$, this can only be because $\varepsilon(k'-1) = \varepsilon(k')$, and again $\varepsilon(k'-1) < \varepsilon'(k-1)$. Repeating the argument, we see that ε never equals ε' to the left of k, a contradiction. □

One more cohomology computation will complete the proof of (8.8). Again, the Cartier isomorphims plays the key role.

8.13 Lemma. Say that ε is "steep at j" iff $\varepsilon(j-1) > \varepsilon(j) > \varepsilon(j+1)$. Then if ε is steep at j and at $j+1$, there is a commutative diagram:

$$\begin{array}{ccc} H^j([\Omega^{\cdot}_{Y'/S}]_\varepsilon \otimes \mathbb{Z}/p\mathbb{Z}) & \xrightarrow{\Psi_\varepsilon \otimes \mathrm{id}} & H^j(F_{Y/S_*}[\Omega^{\cdot}_{Y'/S}]_\eta \otimes \mathbb{Z}/p\mathbb{Z}) \\ \downarrow \cong & & \downarrow \cong \\ \Omega^j_{Y'_0/S_0} & \xrightarrow{C^{-1}} & H^j(F_{Y_0/S_{0*}}\Omega^{\cdot}_{Y_0/S_0}) \end{array}$$

Proof. Set $e = \varepsilon(j)$ and $n = e+j$. Then $\varepsilon(j+i) = e-i$ for $i = -1,0,1,2$ and $\eta(j+i) = n$ for these values of i. The groups $(\Omega^{j+i})_\varepsilon$ and $(\Omega^{j+i})_\eta$ then become relatively simple, and in fact the morphism of complexes $\Psi_\varepsilon \otimes \mathrm{id}$ looks like:

$$\begin{array}{ccccc} p^{e+1}\Omega^{j-1}_{Y'/S}/p^{e+2}\Omega^{j-1}_{Y'/S} & \longrightarrow & p^e\Omega^j_{Y'/S}/p^{e+1}\Omega^j_{Y'/S} & \longrightarrow & p^{e-1}\Omega^{j+1}_{Y'/S}/p^e\Omega^{j+1}_{Y'/S} \\ \downarrow F^*_{Y/S} & & \downarrow F^*_{Y/S} & & \downarrow F^*_{Y/S} \\ F_{Y/S_*}(p^n\Omega^{j-1}_{Y/S}/p^{n+1}\Omega^j_{Y/S}) & \longrightarrow & F_{Y/S_*}(p^n\Omega^j_{Y/S}/p^{n+1}\Omega^j_{Y/S}) & \longrightarrow & F_{Y/S_*}(p^n\Omega^{j+1}_{Y/S}/p^{n+1}\Omega^{j+1}_{Y/S}) \end{array}$$

It is clear that the boundary maps in the upper complex are zero. Thanks to this, we can use division by various powers of p to map the above isomorphically to:

$$\begin{array}{ccccc} \Omega^{j-1}_{Y'_0/S_0} & \xrightarrow{0} & \Omega^j_{Y'_0/S_0} & \xrightarrow{0} & \Omega^{j+1}_{Y'_0/S_0} \\ \downarrow "p^{-j+1}F^*_{Y/S}" & & \downarrow "p^{-j}F^*_{Y/S}" & & \downarrow "p^{-j-1}F^*_{Y/S}" \\ F_{Y/S_*}\Omega^{j-1}_{Y_0/S_0} & \xrightarrow{d} & F_{Y/S_*}\Omega^j_{Y_0/S_0} & \xrightarrow{d} & F_{Y/S_*}\Omega^{j+1}_{Y_0/S_0} \end{array} .$$

Passing to cohomology and invoking (8.5) then proves the lemma. □

8.14 Lemma. Let $f: A^{\bullet} \to B^{\bullet}$ be a map of p-adically separated and complete p-torsion free complexes. If $f \otimes \mathbb{Z}/p\mathbb{Z}$ is a quasi-isomorphism, so is f.

Proof. By passing to mapping cones, it suffices to prove that if a complex $C^{\bullet}$ as above is acyclic mod p, then it is acyclic, i.e. if $A \xrightarrow{\alpha} B \xrightarrow{\beta} C$ is a zero-sequence which is exact mod p, then it is exact. Suppose $\beta(b) = 0$. Then $b = \alpha(a_0) + pb_1$ for some $a_0 \in A$, $b_1 \in B$; and $\beta b_1 = 0$ by the torsion freeness of C. But then $b_1 = \alpha(a_1) + pb_2$, and $b = \alpha(a_0 + pa_1) + p^2 b_2$. By induction, we find elements such that
$b = \alpha(a_0 + pa_1 + \dots p^n a_n) + p^{n+1} b_n$ for all n. Then if $a = \sum_0^{\infty} p^i a_i$, $\alpha(a) = b$. □

We can now complete the proof of Theorem (8.8), for an arbitrary gauge ε. Since the value of $\varepsilon(i)$ for negative i has no effect, we may assume that $\varepsilon(i) = \varepsilon(0) - i$ for all $i \geq 0$; then we can find a gauge $\varepsilon' \geq \varepsilon$ such that $\varepsilon'(i) = \varepsilon(i)$ for almost all i and such that ε' is steep at j and $j+1$ whenever $0 \leq j \leq \dim(Y/S)$:

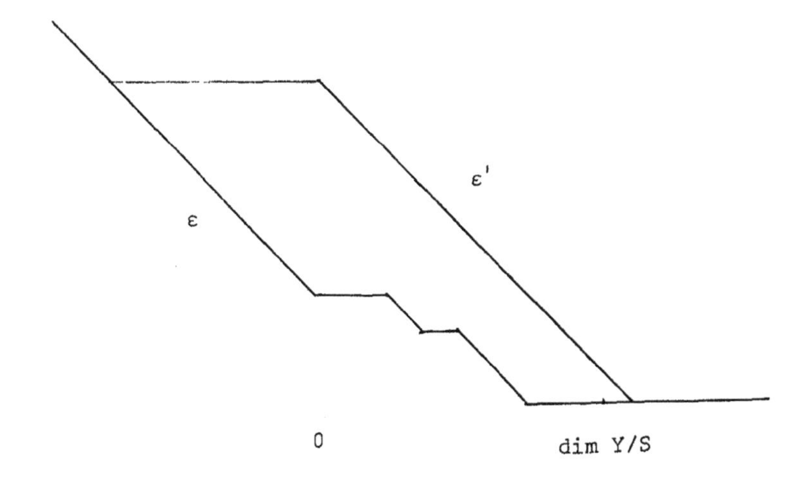

(This would be quite impossible if we were dealing exclusively with tame gauges, and this is the essential reason our proof is simpler than the original one.) It follows from (8.13) that $\Psi_{\varepsilon'} \otimes \mathrm{id}_{\mathbb{Z}/p\mathbb{Z}}$ is a quasi-isomorphism, hence from (8.14) that $\Psi_{\varepsilon'}$ is a quasi-isomorphism. Then Lemma (8.11) implies that Ψ_{ε} is a quasi-isomorphism, and the proof of Theorem (8.8) is therefore complete. □

Our next task is to use crystalline cohomology to find a formulation of (8.8) which has a global meaning — independent of the lifting Y of Y_0 and (more importantly) the lifting F_Y of F_{Y_0}.

8.15 Definition. Let $\mathcal{A}$ be an abelian category, and let p be a fixed prime number. For any filtered object (A,F) of $\mathcal{A}$ and any function $\varepsilon \colon \mathbb{Z} \to \mathbb{N}$, we set:

$$F^{\varepsilon}A = \sum_{i \in \mathbb{Z}} p^{\varepsilon(i)} F^{i}(A) \quad .$$

8.15.1 Remarks. If $F^{\bullet}$ is a decreasing (= nonincreasing) filtration on A, we use the above notation. If $F_{\bullet}$ is an increasing filtration, we use the same definition but write $F_{\varepsilon}A$ instead $F^{\varepsilon}A$. Notice that if F is decreasing, we may as well assume that ε is also decreasing (= nonincreasing) because $F^{\mu}A = F^{\varepsilon}A$, where μ is the maximal decreasing function less than ε, given by $\mu(i) = \min\{\varepsilon(i') \colon i' \leq i\}$. Similarly, if $F_{\bullet}$ is increasing, $F_{\varepsilon}A = F_{\nu}A$, where ν is the maximal increasing function less than ε, given by $\nu(i) = \min\{\varepsilon(i') \colon i' \geq i\}$.

8.15.2 It is clear that if $h: (A,F) \longrightarrow (B,G)$ is a morphism of filtered objects, then h induces a functorial map $h_\varepsilon : F^\varepsilon A \to F^\varepsilon B$. If $\varepsilon \leq \varepsilon'$, there is a natural inclusion $F^{\varepsilon'}A \to F^\varepsilon A$, compatible with the h_ε's.

8.15.3 If $\varepsilon(m)$ is the minimum value of ε, and if $F^\bullet$ is decreasing (resp., increasing), then $F^\varepsilon A = \sum_{-\infty}^{m} P^{\varepsilon(i)} F^i$ (resp., $F_\varepsilon A = \sum_{m}^{\infty} P^{\varepsilon(i)} F_i$). Consequently, if $F^\bullet$ is bounded below (resp., if $F_\bullet$ is bounded above), we only need a finite sum to define $F^\varepsilon A$ (resp., $F_\varepsilon A$).

8.15.4 We shall be especially interested in the case where A is the abelian category of complexes of an abelian category. An important point is that the formation of F^ε is compatible with filtered homotopies. Specifically, if f and g are morphisms of filtered complexes $(A^\bullet,F) \longrightarrow (B^\bullet,G)$, and if R is a filtered homotopy $f \sim g$, then R induces, in the obvious way, a homotopy R_ε between the morphisms $f_\varepsilon, g_\varepsilon$: $F^\varepsilon A^\bullet \to F^\varepsilon B^\bullet$. It seems reasonable to attempt to define the derived functors of the ε-construction, in the context of the filtered derived category. However, we shall not need this.

8.15.5 Let $F^\bullet$ be the "filtration bête" [1, 1.4.7] (the Hodge filtration) on the complex $K^\bullet$, given by $F^iK^j = [0$ if $i > j$, K^j if $i \leq j]$. Then if ε is decreasing, one sees immediately that $F^\varepsilon K^\bullet = K^\bullet_\varepsilon$, in the notation (8.6).

8.15.6 Let $C.$ be the increasing filtration on the complex $K^\cdot$ given by: $C_iK^j = [K^j$ if $j \leq i$, 0 if $j > i+1$, and $\mathrm{Im}(d^i)$ if $j = i+1]$. It is easy to see that if $C^!_\cdot$ is Deligne's "filtration canonique" [1, 1.4.6], there is a canonical filtered quasi-isomorphism $C^!_\cdot K^\cdot \to C.K^\cdot$. Moreover, if η is increasing, there is a canonical quasi-isomorphism: $C_\eta K^\cdot \to (K^\cdot)_\eta$, in the notation of (8.6). We shall not need this remark. □

Assume now that X is a smooth S_0-scheme to which γ extends. Consider the topos $(X/S)_{cris}$ described in (7.17) (recall that we are writing S for $\hat{S}$), and regard the PD filtration $J^{[\cdot]}_{X/S}$ as a filtration on the object $\mathcal{O}_{X/S}$ (we set $J^{[i]}_{X/S} = \mathcal{O}_{X/S}$ if $i \leq 0$). Then according to (8.15), we have

$J^{[\varepsilon]}_{X/S} = \sum_i p^{\varepsilon(i)} J^{[i]}_{X/S}$; note that a finite sum is all we need, by (8.15.3).

8.16 <u>Theorem.</u> Suppose X is closed in Y, Y/S is formally smooth, and $D = \hat{D}_{X,\gamma}(Y)$, the p-adic completion of the PD envelope of X in Y. Then for any $\varepsilon\colon \mathbf{Z} \to \mathbb{N}$, there is a natural quasi-isomorphism:

$$\mathbb{R}u_{X/S_*} J^{[\varepsilon]}_{X/S} \to F^{\varepsilon}_X \Omega^\cdot_{D/S} \quad .$$

<u>Proof.</u> Begin by observing that we can work with $(X/S.)$ instead of (X/S). Indeed, if $j\colon (X/S.)_{cris} \to (X/S)_{cris}$ is the morphism (7.17), one sees immediately that $j^* J^{[\varepsilon]}_{X/S} \cong J^{[\varepsilon]}_{X/S.}$, and that $\mathbb{R}j_* J^{[\varepsilon]}_{X/S.} \cong j_* J^{[\varepsilon]}_{X/S.} \cong J^{[\varepsilon]}_{X/S}$, just as in the proof of (7.19). It follows that there is a canonical isomorphism:

$$\mathbb{R}u_{X/S_*} J^{[\varepsilon]}_{X/S} \cong \mathbb{R}u_{X/S.} J^{[\varepsilon]}_{X/S.} \quad .$$

We now proceed to copy the proof of (7.23). Let us use the notations of that proof. We have a morphism of filtered complexes in (X/S.) $\mu^{\cdot}: J^{[\cdot]}_{X/S.} \to F^{\cdot}_X L(\Omega^{\cdot}_{Y/S.})$, and hence for any ε, a morphism of complexes: $\mu^{[\varepsilon]}: J^{[\varepsilon]}_{X/S.} \to F^{\varepsilon}_X L(\Omega^{\cdot}_{Y/S.})$. To prove that it's a quasi-isomorphism, we may restrict to $(X/S_n)_{cris}$. Now look again at the proof of the filtered Poincaré lemma (6.13). We showed that locally on $Cris(X/S_n)$, the map $\mu^{\cdot}$ was in fact a homotopy equivalence of filtered complexes. It follows from (8.15.4) that $\mu^{[\varepsilon]}$ is also a local homotopy equivalence, and hence a quasi-isomorphism. Consequently, we deduce an isomorphism:

$$\mathbb{R}v_{X/S.*}J^{[\varepsilon]}_{X/S.} \to \mathbb{R}v_{X/S.*}F^{\varepsilon}_X L(\Omega^{\cdot}_{Y/S.}) \quad .$$

Now recall that the qth term of $F^{\varepsilon}_X L(\Omega^{\cdot}_{Y/S})$ is $\Sigma p^{\varepsilon(i)} F^i_X L(\Omega^q_{Y/S.}) = \Sigma p^{\varepsilon(i)} j_{\tilde{Y}*}(J^{[i-q]}\varphi.^*\Omega^q_{Y/S.})$. Since $j_{\tilde{Y}*}$ is exact (5.27), it preserves images and finite sums, and hence this is the same as $j_{\tilde{Y}*}(\Sigma p^{\varepsilon(i)} J^{[i-q]}\varphi.^*\Omega^q_{Y/S.})$. It follows from (7.22.2) that these sheaves are acyclic for $v_{X/S.*}$, so that:

$$\mathbb{R}v_{X/S.*}J^{[\varepsilon]}_{X/S.} \cong (v_{X/S.*} \circ j_{\tilde{Y}*} = \varphi._*)\,(\Sigma p^{\varepsilon(i)} J^{[i-q]}\varphi.^*\Omega^{\cdot}_{Y/S.})$$

But (7.22.1) tells us that $\varphi._*$ is also exact, so the qth term of this complex is

$$\Sigma p^{\varepsilon(i)} \varphi._*(J^{[i-q]}\varphi.^*\Omega^q_{Y/S.}) = \Sigma p^{\varepsilon(i)} J^{[i-q]}_{D.}\Omega^q_{D./S.} \quad .$$

It is clear that these sheaves satisfy the hypotheses of (7.20), hence are acyclic for $\pi_X\colon \widetilde{\mathbb{M}} \times X \to X$. We conclude that:

$$\mathbb{R}u_{X/S.*}J^{[\varepsilon]}_{X/S.} \cong \pi_{X*}(F^{\varepsilon}_X\Omega^{\bullet}_{D./S.}) \quad .$$

There is just one more unpleasant point to check, that $\pi_{X*}(F^{\varepsilon}_X\Omega^{\bullet}_{D./S.}) \cong F^{\varepsilon}_X\Omega^{\bullet}_{D/S}$. This is not quite obvious, because π_{X*} does not preserve images and sums. It is clear that what we must prove is that the natural map:

$\Sigma p^{\varepsilon(i)}J^{[i]}_D \longrightarrow \varprojlim \Sigma p^{\varepsilon(i)}J^{[i]}_{D_n}$ is an isomorphism.

It is apparent from the explicit bases we gave for PD envelopes (3.32ff) that $\mathcal{D}_n$ is a flat $\mathcal{O}_{S_n}$-module, so that the kernel of $\mathcal{D}_{n+k} \to \mathcal{D}_n$ is exactly $p^{n+1}\mathcal{D}_{n+k}$. Recall that an inverse system is said to be "eventually strict" iff the transition maps are surjective for $n \gg 0$. In the next lemma, we work over an affine open set, and write $\mathcal{D}$ for $\Gamma(\mathcal{D})$, etc.

8.17 Lemma. Suppose $M. \subseteq \mathcal{D}.$ is an eventually strict sub inverse system. Then for any e and i, $M. \cap p^eJ.^{[i]}$ is also eventually strict.

Proof. If M_n is strict for $n \geq n_0$, we'll prove that $M_n \cap p^eJ^{[i]}_n$ is strict for $n \geq \sup(n_0, e+i)$. For such an n, suppose $x \in M_n \cap p^eJ^{[i]}_n$, then we find $y \in M_{n+1}$ and $z \in p^eJ^{[i]}_{n+1}$ both mapping to x. Then $\omega = y-z$ lies in $p^{n+1}\mathcal{D}_{n+1} \subseteq p^e(p)^{[i]}\mathcal{D}_{n+1} \subseteq p^eJ^{[i]}_{n+1}$, and hence $y = \omega+z$ lies in $M_{n+1} \cap p^eJ^{[i]}_{n+1}$ and lifts x, as required. □

If now $M_{\cdot} \subseteq \mathcal{D}_{\cdot}$ is quasi-coherent and strict, it follows from the lemma and (7.20) that $\mathbb{R}^1\pi_{X*}(M_{\cdot} \cap p^e J_{\cdot}^{[i]}) = 0$. From the exact sequence

$$0 \to M_{\cdot} \cap p^e J_{\cdot}^{[i]} \to M_{\cdot} \oplus p^e J_{\cdot}^{[i]} \to M_{\cdot} + p^e J_{\cdot}^{[i]} \to 0$$

we now see that $(\varprojlim M_{\cdot}) + \varprojlim\ p^e J_{\cdot}^{[i]} \cong \varprojlim (M_{\cdot} + p^e J_{\cdot}^{[i]})$. By induction, then, we see that $\sum_i \varprojlim p^{\varepsilon(i)} J_{\cdot}^{[i]} \cong \varprojlim (\Sigma p^{e(i)} J_{\cdot}^{[i]})$. (Recall that a finite sum suffices.)

It remains only to verify that $p^e J^{[i]} \cong \varprojlim p^e J_{\cdot}^{[i]}$. From the exact sequence:

$$0 \to J_n^{[i]} \cap p^{n+1-e}\mathcal{D}_n \to J_n^{[i]} \to p^e J_n^{[i]} \to 0 \quad ,$$

we see that it suffices to check that $R^1 \varprojlim (J_n^{[i]} \cap p^{n+1-e}\mathcal{D}_n) = 0$. Since this inverse system is essentially zero, this is clear. This completes the proof that $\pi_{X*} J_{X/S_{\cdot}}^{[\varepsilon]} \cong J_{X/S}^{[\varepsilon]}$, and hence the proof of the theorem. □

It turns out that only certain $\varepsilon\colon \mathbb{Z} \to \mathbb{N}$ carry useful information.

8.18 Definition. If $k \in \mathbb{Z}$, let $\langle k \rangle = \min\{\mathrm{ord}_p(p^r/r!) : r \geq k\}$, so that $(p)^{[k]} = (p^{\langle k \rangle})$ (= the unit ideal if $k \leq 0$). A map $\varepsilon\colon \mathbb{Z} \longrightarrow \mathbb{N}$ is "tame" iff $\varepsilon(j) - \varepsilon(i) \leq \langle i-j \rangle$ for all i and j .

8.18.1 Remarks. If a and b are integers, $(p)^{[a]}(p)^{[b]} \subseteq (p)^{[a+b]}$, and hence $\langle a+b \rangle \leq \langle a \rangle + \langle b \rangle$. In particular, $\langle k \rangle \leq \langle k+1 \rangle \leq \langle k \rangle + 1$. Notice that if $p = 2$, $\langle k \rangle = 1$ whenever $k \geq 1$. Any tame ε is a gauge (8.7).

8.18.2 Suppose $\varepsilon: \mathbb{Z} \to \mathbb{N}$ is any function, and set $\bar{\varepsilon}(j) = \min\{\varepsilon(k) + \langle k-j\rangle : k \in \mathbb{Z}\}$. Then it is not hard to see that $\bar{\varepsilon}$ is the maximal tame gauge less than or equal to ε. Moreover, we have $J_{X/S}^{[\varepsilon]} = J_{X/S}^{[\bar{\varepsilon}]}$. Indeed, since $\bar{\varepsilon} \leq \varepsilon$, we need only check that $J_{X/S}^{[\bar{\varepsilon}]} \subseteq J_{X/S}^{[\varepsilon]}$, i.e., that each $p^{\bar{\varepsilon}(j)} J_{X/S}^{[j]} \subseteq \sum_k p^{\varepsilon(k)} J_{X/S}^{[k]}$. But if $\bar{\varepsilon}(j) = \varepsilon(k) + \langle k-j\rangle$, $p^{\bar{\varepsilon}(j)} J_{X/S}^{[j]} = p^{\varepsilon(k)}(p)^{[k-j]} J_{X/S}^{[j]} \subseteq p^{\varepsilon(k)} J_{X/S}^{[k]}$, because $p \in J$. This is why we can only consider tame gauges.

8.18.3 For any $i > 0$, set $\varepsilon_i(j) = \langle i-j\rangle$. Then ε_i is tame, and $J_{X/S}^{[i]} = J^{[\varepsilon_i]}_{X/S}$. This gives us a "gauge theoretic" interpretation of the P.D. filtration $J_{X/S}^{[\cdot]}$.

8.18.4 Suppose ε is tame, and Y/S is smooth and $X = Y_0$. Then $\mathbb{R}u_{X/S*} J_{X/S}^{[\varepsilon]} \cong F_X^{\varepsilon}\Omega^{\cdot}_{Y/S} = F_Y^{\varepsilon}\Omega^{\cdot}_{Y/S} = (\Omega^{\cdot}_{Y/S})_{\varepsilon}$. This tells us that the p-adic interpolation of the Hodge filtration Y defined by the tame gauge ε depends only on X, not Y. □

The above remarks provide us with a crystalline interpretation of the source of the arrow Ψ_{ε} of (8.8). The target turns out to be quite easy to handle, for abstract nonsense will tell us that "formation of $K^{\cdot}_{\eta}$" passes over to the derived category, if η is increasing. Notice first that if f and g are homotopic maps $A^{\cdot} \to B^{\cdot}$, then f_{η} and g_{η} are homotopic. Indeed, if $R^q: A^q \to B^{q-1}$ is the homotopy, it follows immediately from the definitions and the fact that η is increasing that R^q

maps A^q_η into B^{q-1}_η, and hence induces a homotopy $R^q_\eta\colon f_\eta \sim g_\eta$. Of course, formation of $K^\bullet_\eta$ is not compatible with translation, hence doesn't preserve triangles, nor is it exact. Nevertheless, it has a left derived functor, in a slightly extended sense, because of the following simple result:

8.19 Proposition. Let $f\colon A^\bullet \to B^\bullet$ be a quasi-isomorphism of sheaves of abelian groups. Assume that $A^\bullet$ and $B^\bullet$ are p-torsion free and that $\eta\colon \mathbb{Z} \to \mathbb{N}$ is increasing. Then $f_\eta\colon A^\bullet_\eta \to B^\bullet_\eta$ is also a quasi-isomorphism.

Proof. We just check the stalks, so we work with groups instead of sheaves. Assuming that $H^i(f)$ is an isomorphism:

(i) $H^i(f_\eta)$ is injective: Any class in $H^i(A^\bullet_\eta)$ is repre- represented by some $p^{\eta(i)}a$, with $da = 0$, since A^{i+1} is torsion free. If $H^i(f_\eta)[a] = 0$, $f(p^{\eta(i)}(a)) = dp^{\eta(i-1)}b'$, for some $b' \in B^{i-1}$. Then $f(p^{\eta(i)-\eta(i-1)}a) = db'$ (this makes sense because η is increasing), and since $H^i(f)$ is injective, $p^{\eta(i)-\eta(i-1)}a = da'$ for some $a' \in A^{i-1}$. Then $p^{\eta(i)}a = dp^{\eta(i-1)}a'$, and $p^{\eta(i-1)}a' \in A^{i-1}_\eta$, so $[p^{\eta(i)}a] = 0$ in $H^i(A^\bullet_\eta)$.

(ii) $H^i(f_\eta)$ is surjective: Any class in $H^i(B^\bullet_\eta)$ is repre- sented by some $p^{\eta(i)}b$, with $b \in B^i$, $db = 0$. Since $H^i(f^\bullet)$ is surjective, we can choose an a in A^i with $da = 0$ and a b' in B^{i-1} such that $b = f(a) + db'$. Then $p^{\eta(i)}b = f(p^{\eta(i)}a) + dp^{\eta(i)}b'$. Certainly

$p^{\eta(i)}a \in A^i_\eta$, and $p^{\eta(i)}b' \in B^{i-1}_\eta$ because η is non-decreasing. Thus $[p^{\eta(i)}b] = H^i(f_\eta)[p^{\eta(i)}a]$. □

To construct the left derived functor $\mathbb{L}\eta$: $D(X) \to D(X)$, let $F^\cdot$ be a complex of abelian sheaves on X, and let $L^\cdot(F^\cdot) \to F^\cdot$ be a flat resolution. Since $\mathbb{Z}$ has finite projective dimension, this makes sense even if $F^\cdot$ is unbounded, and we can take $L^\cdot(F^\cdot)$ to be bounded below if $F^\cdot$ is. Then define $\mathbb{L}\eta(F^\cdot)$ to be $L^\cdot(F^\cdot)_\eta$. This makes sense in the derived category, thanks to the previous result.

We now can state and prove the crystalline version of the main local theorem:

8.20 Theorem. Let X/S_0 be smooth, with S as in (8.1). If ε is any tame gauge and if $\eta(i) = \max\{0, \varepsilon(i)+i\}$ for all i, there is a commutative diagram as shown, in which Ψ_ε is an isomorphism.

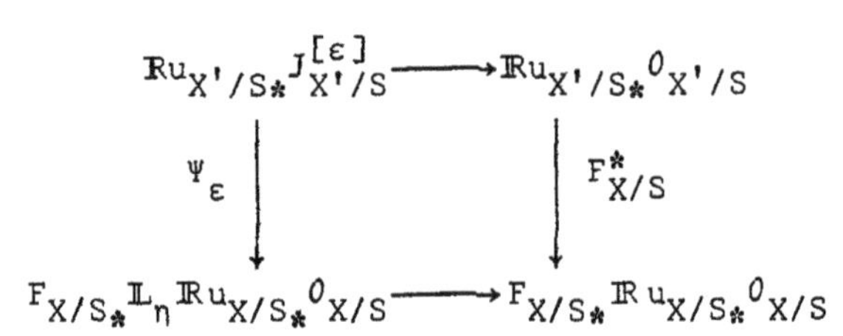

$$\begin{array}{ccc} \mathbb{R}u_{X'/S*}J^{[\varepsilon]}_{X'/S} & \longrightarrow & \mathbb{R}u_{X'/S*}\mathcal{O}_{X'/S} \\ \downarrow \Psi_\varepsilon & & \downarrow F^*_{X/S} \\ F_{X/S*}\mathbb{L}_\eta\mathbb{R}u_{X/S*}\mathcal{O}_{X/S} & \longrightarrow & F_{X/S*}\mathbb{R}u_{X/S*}\mathcal{O}_{X/S} \end{array}$$

This diagram is functorial in X and ε , and agrees with the diagram (8.8) in a lifted situation.

Proof. The meaning of the horizontal arrows is clear. The main remaining point is the existence of the arrow Ψ_ε .

It would be nice to have an intrinsic proof; we have to resort to an unpleasant local calculation.

8.21 <u>Lemma.</u> Suppose $(Y/S, F_Y)$ is a lifted situation, and $X \hookrightarrow Y$ is a closed S_0-scheme. Let $D = \hat{D}_{X,Y}(Y)$, $D' = \hat{D}_{X',Y}(Y')$. Then $F_{Y/S}$ induces a natural morphism of complexes: $\Psi_{D/S}\colon \Omega^{\cdot}_{D'/S} \to F_{X/S*}\Omega^{\cdot}_{D/S}$. Moreover:

8.21.1 For each i, $\Psi_{D/S}$ maps $F^i_{X'}\Omega^{\cdot}_{D'/S}$ into $F_{X/S*}(\Omega^{\cdot}_{D/S})_{\zeta_i}$, where

$$\zeta_i(j) = \max\{0, [\langle i-j\rangle + j]\} \quad .$$

8.21.2 For any gauge ε, $\Psi_{D/S}$ maps $F^{\varepsilon}_{X'}\Omega^{\cdot}_{D'/S}$ into $F_{X/S*}(\Omega^{\cdot}_{D/S})_{\zeta}$, where

$$\zeta(j) = \max\{0, \min\{\varepsilon(i) + j + \langle i-j\rangle : i \in \mathbb{Z}\}\} .$$

<u>Proof.</u> Look at the relative Frobenius diagram (8.2.1). Since X is an S_0-scheme and $p\mathcal{O}_{S_0} = 0$, F_Y maps X to X, inducing $F_X\colon X \to X$. Moreover, $W_{Y/S}$ maps X' to X and $F_{Y/S}$ maps X to X', inducing $F_{X/S}$, and $X' = W^{-1}_{Y/S}(X)$. It follows that we get induced maps $W_{D/S}$ and $F_{D/S}$ on the PD envelopes, and $X' = W^{-1}_{D/S}(X)$. Now $D_0 = D_{X,Y}(Y_0)$ is in characteristic $p > 0$, and $F_{D_0} = W_{D_0/S_0} \circ F_{D_0/S_0}$ is its absolute Frobenius morphism. Since the ideal J_{D_0} of X in $\mathcal{D}_0$ is a PD ideal, $F^*_{D_0}(J_{D_0}) = 0$, hence also $F^*_{D_0/S_0}(J_{D_0'}) = 0$. Since $D_0 = D\times_S S_0$, this implies that $F^*_{D/S}(J_{D'}) \subseteq (p)\, F_{D/S*}(\mathcal{D})$, and since $F^*_{D/S}$ is a

PD-morphism, it follows that $F^*_{D/S}(J^{[i]}_{D'}) \subseteq (p)^{[i]}F_{D/S*}(\mathcal{D})$ for all i.

Now as we observed in the lifted situation (8.2), $\Psi_{Y/S}$ maps $\Omega^j_{Y/S}$ into $p^jF_{Y/S*}(\Omega^j_{Y/S})$, so the same is true for $\Psi_{D/S}$. Since $F^i_X\Omega^j_{D'/S}$ is by definition $J^{[i-j]}_{D'}\Omega^j_{D'/S}$, it is clearly mapped by $\Psi_{D/S}$ into $(p)^{<i-j>+j}F_{D/S_*}(\mathcal{D}\otimes\Omega^j_{D/S})$. Since $\Psi_{D/S}$ is a morphism of complexes, (8.21.1) follows, and (8.21.2) is an immediate consequences. $\square$

Now if ε is <u>tame</u>, $\varepsilon(j) \leq \varepsilon(i) + <i-j>$ for all i, hence $\zeta(j) = \varepsilon(j) + j = \eta(j)$, if $j > 0$. Because the complex $\Omega^{\cdot}_{D/S}$ is p-torsion free, we see that $\Psi_{D/S}$ induces the desired arrow Ψ_ε.

It is important to note that if $X \longrightarrow Z$ is another closed immersion into a lifted situation, and if we can find a morphism $(Y,F_Y) \to (Z,F_Z)$ compatible with the inclusions of X, then the morphisms of complexes we have constructed are compatible, in the evident sense. That is, there is a commutative diagram:

$$\begin{array}{ccc} F^\varepsilon_{X'}\Omega^{\cdot}_{D_{X',\gamma}(Z)/S} & \longrightarrow & F_{X/S*}(\Omega^{\cdot}_{D_{X,\gamma}(Z)/S})_\eta \\ \downarrow & & \downarrow \\ F^\varepsilon_{X'}\Omega^{\cdot}_{D_{X',\gamma}(Y)/S} & \longrightarrow & F_{X/S*}(\Omega^{\cdot}_{D_{X,\gamma}(Y)/S})_\eta \end{array}$$

Moreover, if (Y,F_Y) and (Z,F_Z) are any two lifted situations in which X embeds, then we can also embed X in $(Y \times Z,\ F_Y \times F_Z)$ (as a <u>locally</u> closed subscheme, but no matter) — and this maps both to (Y,F_Y) and to (Z,F_Z). It follows that the arrow $\Psi_\varepsilon\colon \mathbb{R}u_{X'/S*}J^{[\varepsilon]}_{X/S} \to F_{X/S*}\,\mathbb{L}\eta\mathbb{R}u_{X/S*}\mathcal{O}_{X/S}$ in the derived category that we have defined is independent of the choice of embedding.

Let us observe that the theorem is now proved for quasi-projective X/S. Indeed, the absolute Frobenius of $\mathbb{P}_{S_0}$ lifts to $\mathbb{P}_S$, so X can be embedded in a lifted situation. Moreover, once the arrow Ψ_ε is defined, it must be a quasi-isomorphism, because this is a local question, and we may therefore assume that (X,F_X) lifts. Because the arrow Ψ_ε is independent of the choice of embedding, we can use the lifting to calculate it, so that (8.8) implies that it is a quasi-isomorphism.

It remains only to use cohomological descent to define the arrow in the general case, as Deligne suggests. We can easily find an open covering $X^0 \to X$ such that $X^0 \hookrightarrow Y^0$ with (Y^0, F_{Y^0}) a lifted situation: Then $X^n = X^0\times_X\ldots X^0 \hookrightarrow Y^0\times_S Y^0\ldots Y^0 = Y^n$ is locally closed, and $(Y^n, F_Y\times_S\ldots F_Y)$ is a lifted situation. Because each $\pi_n\colon X^n \to X$ is locally an open immersion, $\pi_n^*\,\mathbb{R}u_{X/S*}F \cong \mathbb{R}u_{X^n/S*}\pi^*_{n\,\mathrm{cris}}F$ for any abelian $F \in (X/S)_{\mathrm{cris}}$. We now have natural isomorphisms (using the notation of (7.8)):

$$\mathbb{R}u_{X'/S*}J^{[\varepsilon]}_{X'/S} \cong \mathbb{R}\pi_*\pi^*\,\mathbb{R}u_{X'/S*}J^{[\varepsilon]}_{X'/S} \cong \mathbb{R}\pi_*\,\mathbb{R}u_{X'^{\bullet}/S*}J^{[\varepsilon]}_{X'^{\bullet}/S}$$

$$\overset{\Psi_\varepsilon}{\cong} \mathbb{R}\pi_*\,\mathbb{L}\eta\mathbb{R}u_{X^{\bullet}/S*}\mathcal{O}_{X^{\bullet}/S} \cong \mathbb{R}\pi_*\,\mathbb{L}\eta\,\pi^*\,\mathbb{R}u_{X/S*}\mathcal{O}_{X/S}$$

$$\cong \mathbb{R}\pi_*\pi^*\,\mathbb{L}\eta\,\mathbb{R}u_{X/S*}\mathcal{O}_{X/S} \cong \mathbb{L}\eta\,\mathbb{R}u_{X/S*}\mathcal{O}_{X/S}. \qquad \square$$

In order to draw the cohomological consequences of the main local theorem (8.20), it is convenient to know some of the properties of the functors $\mathbb{L}\eta$. It is also convenient to make a slight additional restriction on η .

8.22 <u>Definition.</u> A function $\eta\colon \mathbb{Z} \to \mathbb{N}$ is a "cogauge" iff

$$\eta(i) \le \eta(i+1) \le \eta(i)+1 \quad \text{for all } i .$$

8.22.1 <u>Remarks.</u> Suppose that $K^\cdot$ is a p-torsion free complex and η is a cogauge. Then multiplication by p^k induces an isomorphism: $K^\cdot_\eta \to K^\cdot_{\eta+k}$. This translates into a statement in the derived category: There is a commutative diagram (not a triangle!):

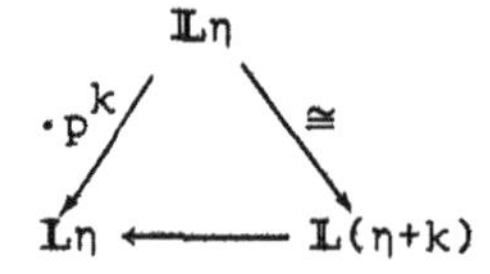

We call this the "shifting diagram".

8.22.2 If η and ζ are cogauges, so are $\eta \vee \zeta = \max\{\eta,\zeta\}$ and $\eta \wedge \zeta = \min\{\eta,\zeta\}$, (but $\eta+\zeta$ need not be). If $K^\cdot$ is torsion free, it is easy to verify that $K^\cdot_\eta + K^\cdot_\zeta = K^\cdot_{\eta\wedge\zeta}$, $K^\cdot_\eta \cap K^\cdot_\zeta = K^\cdot_{\eta\vee\zeta}$, and $(K^\cdot_\eta)_\zeta = K^\cdot_{\eta+\zeta}$. (The first of these requires the cogauge condition.) These statements translate in various ways into the derived category, which the reader can imagine.

8.23.3 If $\eta \le \eta'$, there is a natural exact sequence: $0 \to K^\cdot_{\eta'} \to K^\cdot_\eta \to K^\cdot_\eta/K^\cdot_{\eta'} \to 0$. In the derived category, this translates into a triangle:

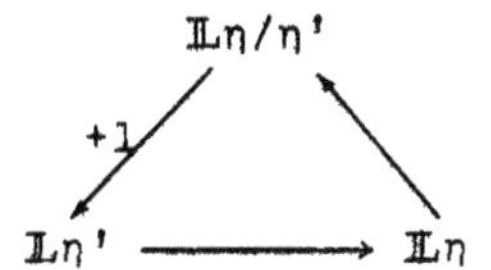

where $\mathbb{L}\eta/\eta'$ is the mapping cone of $\mathbb{L}\eta' \to \mathbb{L}\eta$.

A diagram of inclusions:

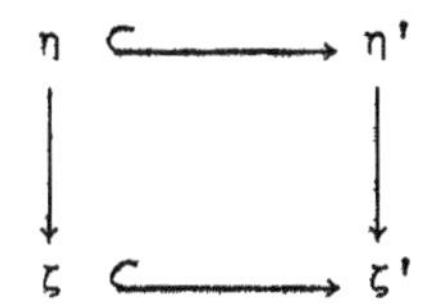

induces a morphism of triangles, which we prefer to notate as short exact sequences:

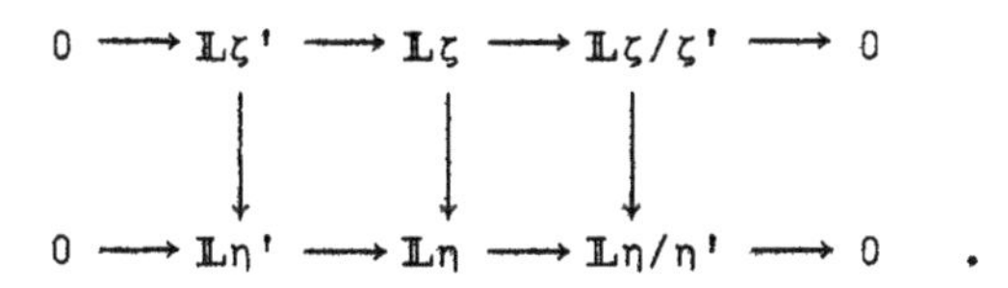

These diagrams are functorial, compatible with shifts, etc. A convenient way to express the content of the previous remark is to observe:

8.23.4 If $\eta' \wedge \zeta = \eta$ and $\eta' \vee \zeta = \zeta'$, then the canonical map $\mathbb{L}\zeta/\zeta' \to \mathbb{L}\eta/\eta'$ is an isomorphism. It suffices to check this for torsion free complexes, with the triangles replaced by short exact sequences. Then in fact $K^{\bullet}_{\zeta}/K^{\bullet}_{\zeta'} \to K^{\bullet}_{\eta}/K^{\bullet}_{\eta'}$ is an isomorphism as the reader can easily verify.

8.23.5 Suppose that $r = \max\{i: \eta(i) = 0\}$, with $r = \pm\infty$ allowed. We call r the "turn-on" value of η. If $K^\cdot$ is p-torsion free, there is an exact sequence:

$$0 \to pK^\cdot \cap K^\cdot_\eta \to K^\cdot_\eta \to T_r(K^\cdot \otimes \mathbb{Z}/p\mathbb{Z}) \to 0,$$

where $T.$ is the "canonical" filtration, given by:

$$T_r A^\cdot = \ldots \to A^{q-1} \to A^q \to \ldots A^{r-1} \to Z^r \to 0 \to \ldots$$

This statement passes over to the derived category. Indeed, formation of T_r preserves homotopies and quasi-isomorphisms, hence passes over to the derived category, and $pK^\cdot \cap K^\cdot_\eta = K^\cdot_{\eta \vee 1}$. Thus, we have a canonical triangle:

$$0 \to \mathbb{L}_{\eta \vee 1} \to \mathbb{L}\eta \to T_r \circ \overset{\mathbb{L}}{\otimes} id_{\mathbb{Z}/p\mathbb{Z}} \to 0 .$$

8.23.6 Suppose that η' is a simple augmentation of η at j. Then a duplication of the calculation (8.10.1) shows that there is a natural triangle:

$$0 \to \mathbb{L}\eta' \to \mathbb{L}\eta \to H^j(\circ \overset{\mathbb{L}}{\otimes} id_{\mathbb{Z}/p\mathbb{Z}})[-j] \to 0 .$$

Formation of these triangles is natural, compatible with shifts, etc.[1] □

If $K^\cdot$ is a bounded complex of sheaves on X, the assignment $\eta \to \mathbf{H}^*(X, \mathbb{L}\eta K^\cdot)$ defines a rather intricate structure — essentially the "gauge structure" of Mazur [5]. I do not know much about its meaning — perhaps it is correct to think of it as a p-adic elaboration of the "conjugate filtration" of hypercohomology.

[1]The reader who so desires can now skip to Katz's conjecture, p.8.42.

Here we shall develop only those properties of this structure we need for the applications, and refer the reader to Mazur's papers for more details. Here is an important special case:

8.24 Proposition. Suppose that $K^{\cdot}$ is a bounded complex of abelian sheaves on X such that:

(a) $\mathbb{H}^{*}(X,K^{\cdot})$ is p-torsion free.

(b) The spectral sequence:

$$E_2^{s,t} = H^s(X,\mathcal{H}^t(K^{\cdot}\overset{\mathbb{L}}{\otimes}\mathbb{Z}/p\mathbb{Z})) \Rightarrow \mathbb{H}^{*}(X,K^{\cdot}\overset{\mathbb{L}}{\otimes}\mathbb{Z}/p\mathbb{Z}) \text{ is degenerate at } E_2.$$

Then for any two cogauges η and ζ,

8.24.1 If $\eta \geq \zeta$, the map $\mathbb{H}^{*}(X,\mathbb{L}\eta K^{\cdot}) \to \mathbb{H}^{*}(X,\mathbb{L}\zeta K^{\cdot})$ is injective.

8.24.2 If H_{η}^{*} (etc.) is the image of $H^{*}(X,\mathbb{L}\eta K^{\cdot}) \hookrightarrow \mathbb{H}^{*}(X,K^{\cdot})$, we have $H_{\eta}^{*} + H_{\zeta}^{*} = H_{\eta\wedge\zeta}^{*}$ and $H_{\eta}^{*} \cap H_{\zeta}^{*} = H_{\eta\vee\zeta}^{*}$.

Proof. Replace $K^{\cdot}$ by a flat complex $L^{\cdot}$ (which is still bounded), so that $\mathbb{H}^{*}(X,\mathbb{L}\eta K^{\cdot}) \cong \mathbb{H}^{*}(X,L_{\eta}^{\cdot})$ for any η. The hypotheses (a) and (b) hold with $L^{\cdot}$ in place of $K^{\cdot}$. Thus, we may as well assume that $K^{\cdot}$ itself is torsion free.

We prove (8.24.1) by studying a succession of cases.

Case 1. Let $\sigma_r(i) = \{0 \text{ if } i \leq r,\ 1 \text{ if } i > r\}$. Then the map $\mathbb{H}^{*}(K_{\sigma_{r-1}}^{\cdot}) \to \mathbb{H}^{*}(K_{\sigma_r}^{\cdot})$ is injective.

To prove this, observe that σ_{r-1} is a simple augmentation of σ_r at r, so we have commutative diagrams:

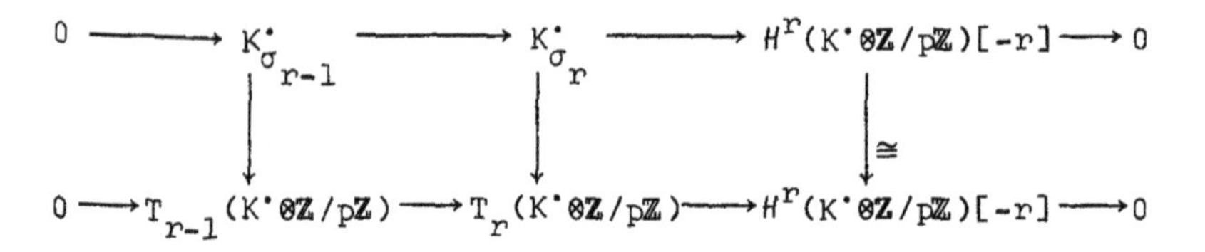

$$\begin{array}{ccccccc}
H^{i-r-1}(\mathcal{H}^r) & \longrightarrow & H^i(K^\cdot_{\sigma_{r-1}}) & \xrightarrow{\beta} & H^i(K^\cdot_{\sigma_r}) & \xrightarrow{\gamma} & H^{i-r}(\mathcal{H}^r) \\
\downarrow{\cong} & & \downarrow{\alpha_{r-1}} & & \downarrow{\alpha_r} & & \downarrow{\cong} \\
H^{i-r-1}(\mathcal{H}^r) & \longrightarrow & H^i(T_{r-1}) & \xrightarrow{\bar{\beta}} & H^i(T_r) & \xrightarrow{\bar{\gamma}} & H^{i-r}(\mathcal{H}^r)
\end{array}$$

First of all, because $K^\cdot$ is bounded, the maps $K^\cdot_{\sigma_r} \to K^\cdot$ and $T_r(K^\cdot \otimes \mathbf{Z}/p\mathbf{Z}) \to K^\cdot \otimes \mathbf{Z}/p\mathbf{Z}$ are isomorphisms for $r \gg 0$, and assumption (a) implies that $\mathbb{H}^i(K^\cdot) \to \mathbb{H}^i(K^\cdot \otimes \mathbf{Z}/p\mathbf{Z})$ is surjective for all i. Therefore α_r is surjective for $r \gg 0$. Descending induction on r, using the above diagram, implies that α_r is surjective for all r. Assumption (b) implies that $\bar{\gamma}$ is surjective, and the diagram implies that γ is surjective. Case 1 follows.

Because $K^\cdot$ is bounded above, the following is true for all $s \gg 0$:

L(s): The map $\mathbb{H}^*(K_\eta) \to \mathbb{H}^*(K_\zeta)$ is injective if $\eta(i) = \zeta(i)$ for all $i \leq s$.

Because $K^\cdot$ is bounded below, it suffices to prove that L(s) is true for all s, and we can use descending induction on s.

<u>Case 2.</u> Let $\zeta_r(i) = \{0 \text{ if } i \leq r,\ i-r \text{ if } i \geq r\}$. Then if L(r) holds, the map $\mathbb{H}^*(K^\cdot_{\zeta_r'}) \to \mathbb{H}^*(K^\cdot_{\zeta_r})$ is injective, where ζ_r' is the simple augmentation of ζ_r at r.

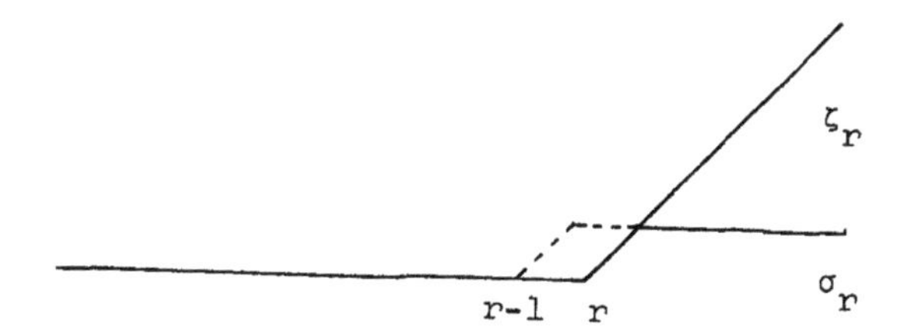

We have commutative squares:

$$\begin{array}{ccc} \zeta'_r & \geq & \zeta_r \\ \vee\!| & & \vee\!| \\ \sigma_{r-1} & \geq & \sigma_r \end{array} \qquad \begin{array}{ccc} \mathbb{H}^*(K^{\bullet}_{\zeta'_r}) & \xrightarrow{a} & \mathbb{H}^*(K^{\bullet}_{\zeta_r}) \\ \downarrow b & & \downarrow d \\ \mathbb{H}^*(K^{\bullet}_{\sigma_{r-1}}) & \xrightarrow{c} & \mathbb{H}^*(K^{\bullet}_{\sigma_r}) \end{array} .$$

The assumption $L(r)$ implies that b is injective, and Case 1 implies that c is injective. Therefore a is injective.

<u>Case 3.</u> If η is a simple augmentation of ζ at $r' \geq r$, and if $L(r)$ holds, then the map $\mathbb{H}^*(K^{\bullet}_{\eta}) \to \mathbb{H}^*(K^{\bullet}_{\zeta})$ is injective.

<u>Proof.</u> First of all, shifting Case 2 implies that the map $\mathbb{H}^*(K^{\bullet}_{\zeta'_r+n}) \to \mathbb{H}^*(K^{\bullet}_{\zeta_r+n})$ is injective, where $n = \zeta(r)$. Now if $r' > r$, the claim is trivial, and if $r' = r$ we have:

$$\begin{array}{ccc} \zeta'_r+n & \geq & \zeta_r+n \\ \vee\!| & & \vee\!| \\ \eta & \geq & \zeta \end{array} \; : \; \begin{array}{ccccc} \mathbb{H}^i(K^{\bullet}_{\zeta'_r+n}) & \longrightarrow & \mathbb{H}^i(K^{\bullet}_{\zeta_r+n}) & \xrightarrow{\alpha} & H^{i-r}(\mathcal{H}^r) \\ \downarrow & & \downarrow & & \downarrow \cong \\ \mathbb{H}^i(K^{\bullet}_{\eta}) & \longrightarrow & \mathbb{H}^i(K^{\bullet}_{\zeta}) & \xrightarrow{\beta} & H^{i-r}(\mathcal{H}^r) \end{array} .$$

Now the map α is surjective, hence β is also surjective, and Case 3 follows.

To finish the proof of (8.24.1), we show that $L(r)$ implies $L(r-1)$. Suppose that $\eta \geq \zeta$ and $\eta(i) = \zeta(i)$ if $i \leq r-1$.

We may as well assume that $\eta(i) = \zeta(i)$ for $i >> 0$. But then it is easy to find a chain of simple augmentations $\zeta_0 = \eta_0 \leq \eta_1 \leq \cdots \leq \eta_n = \eta$ (for instance, one can use Lemma 8.12 by considering the gauges ζ-id+M, η-id+M for large M). Of course, all these are simple augmentations at some $r' \geq r$, so by case 3, the maps $\mathbb{H}^*(K^\cdot_{\eta_{i-1}}) \to \mathbb{H}^*(K^\cdot_{\eta_i})$ are all injective.

It is now easy to derive (8.24.2). We have an exact sequence of complexes:

$$0 \to K^\cdot_{\eta\vee\zeta} \to K^\cdot_\eta \oplus K^\cdot_\zeta \to K^\cdot_{\nu\wedge\zeta} \to 0$$

(8.24.1) tells us that this gives us short exact sequences:

$$\begin{array}{ccccccccc} 0 & \longrightarrow & \mathbb{H}^*(K^\cdot_{\eta\vee\zeta}) & \longrightarrow & \mathbb{H}^*(K^\cdot_\eta) \oplus \mathbb{H}^*(K^\cdot_\zeta) & \longrightarrow & \mathbb{H}^*(K^\cdot_{\eta\wedge\zeta}) & \longrightarrow & 0 \\ & & \downarrow\cong & & \downarrow\cong & & \downarrow\cong & & \\ 0 & \longrightarrow & H^*_{\eta\vee\zeta} & \longrightarrow & H^*_\eta \oplus H^*_\zeta & \longrightarrow & H^*_{\eta\wedge\zeta} & \longrightarrow & 0 \end{array} \quad . \quad \square$$

The following result describes the cohomology of $J^{[\varepsilon]}_{X/S}$ in a special case.

8.25 <u>Proposition.</u> Suppose that A is as in (8.1), that $Y/\mathit{Spec}\ A$ is smooth and proper, and that all its Hodge groups $H^q(Y, \Omega^p_{Y/S})$ are p-torsion free. Then if $\varepsilon\colon \mathbb{Z} \to \mathbb{N}$ is any decreasing function, there is a canonical isomorphism: $\mathbb{H}^i(Y, F^\varepsilon\Omega^\cdot_{Y/S}) \to F^\varepsilon\mathbb{H}^i(Y, \Omega^\cdot_{Y/S})$, where F denotes the Hodge filtration.

<u>Proof.</u> Thanks to Grothendieck's fundamental theorem, we can work with the scheme Y rather than its p-adic completion. There is a

natural morphism of filtered complexes: $j\colon (F^{\varepsilon}\Omega^{\cdot}_{Y/S},F^{\cdot}) \to (\Omega^{\cdot}_{Y/S},F^{\cdot})$ and hence a morphism of spectral sequences:
$j^{st}_r\colon E^{st}_r(F^{\varepsilon}\Omega^{\cdot}_{Y/S},F^{\cdot}) \to E^{st}_r(\Omega^{\cdot}_{Y/S},F^{\cdot})$.

Let $S' = S\times_{Sp\,\mathbb{Z}} Sp\,\mathbb{Q}$, and observe that $j\otimes id_{\mathbb{Q}}$ is an isomorphism. Since taking cohomology commutes with flat base change, it follows that $j^{st}_r\otimes id_{\mathbb{Q}}$ is an isomorphism for all s,t,r .

Now the map j^{st}_1 is just the obvious one: $H^t(Y,p^{\varepsilon(s)}\Omega^s_{Y/S}) \to H^t(Y,\Omega^s_{Y/S})$, and since $\Omega^s_{Y/S}$ is p-torsion free, this map is an isomorphism onto $p^{\varepsilon(s)}H^t(Y,\Omega^s_{Y/S})$. In particular, $E^{st}_1(F^{\varepsilon}\Omega^{\cdot}_{Y/S})$ is also p-torsion free for all s,t . Furthermore, Hodge theory tells us that $d^{st}_r(\Omega^{\cdot}_{Y/S})\otimes id_{\mathbb{Q}} = d^{st}_r(\Omega^{\cdot}_{Y'/S'}) = 0$ for all s,t,r. By induction on r we see that $E^{st}_r(\Omega^{\cdot}_{Y/S})$ and $E^{st}_r(F^{\varepsilon}\Omega^{\cdot}_{Y/S})$ are torsion free and that $d^{st}_r(\Omega^{\cdot}_{Y/S}) = d^{st}_r(\Omega^{\cdot}_{Y'/S'}) = 0$ for all r .

Thus, the map $j^i\colon F^{\cdot}\mathbb{H}^i(F^{\varepsilon}\Omega^{\cdot}_{Y/S}) \to F^{\cdot}\mathbb{H}^i(\Omega^{\cdot}_{Y/S})$ induces an isomorphism:

$$gr^k_F\mathbb{H}^i(F^{\varepsilon}\Omega^{\cdot}_{Y/S}) \to p^{\varepsilon(k)}gr^k_F\mathbb{H}^i(\Omega^{\cdot}_{Y/S}) \quad .$$

It follows that j^i is injective, and (by induction on s), j^i induces an isomorphism

$$F^{i-s}\,\mathbb{H}^i(F^{\varepsilon}\Omega^{\cdot}_{Y/S}) \to p^{\varepsilon(i)}F^i + p^{\varepsilon(i-1)}F^{i-1}+\ldots+ p^{\varepsilon(i-s)}F^{i-s} \subseteq \mathbb{H}^i(\Omega^{\cdot}_{Y/S}).$$

Taking $s = i$, one has the proposition. □

If ε is tame, (8.18.4) shows that $F^{\varepsilon}\Omega^{\cdot}_{Y/S}$ is $\mathbb{R}u_{Y_0/S*}J^{[\varepsilon]}_{Y_0/S}$, and hence depends only on $X = Y_0$. Moreover, (8.20) tells us that relative Frobenius $F_{Y_0/S}$ induces a map (in the derived category):

$F^{\varepsilon}\Omega^{\cdot}_{Y/S} \to [\Omega^{\cdot}_{Y/S}]_{\eta}$, where $\eta(i) = \varepsilon(i) + i$. Since $(\Omega^{\cdot}_{Y/S})_{\eta} \subseteq p^{\varepsilon(0)}\Omega^{\cdot}_{Y/S}$, this implies that on cohomology, $F_{Y_0/S}$ maps $\mathbb{H}^i(Y',F^{\varepsilon}\Omega^{\cdot}_{Y'/S})$ into $p^{\varepsilon(0)}\,\mathbb{H}^i(Y,\Omega^{\cdot}_{Y/S})$ – provided, of course, that ε is tame. Let me remark that tameness of ε really is essential. Indeed, it is easy to see that if the above "divisibility" held for all ε , then (equivalently, in fact), $F_{Y_0/S}F^k\,\mathbb{H}^i(Y',\Omega^{\cdot}_{Y'/S}) \subseteq p^k\,\mathbb{H}^i(Y,\Omega^{\cdot}_{Y/S})$ for all k – and this is <u>false</u>, in general.

It is nonetheless apparent that there should be some relation between the Hodge filtration and $F_{X/S}$, and from (8.23.5), we can also expect the conjugate filtration $F^{\cdot}_{con}$ (associated to the spectral sequence $E_2^{pq} = H^p(X,\mathcal{H}^q(\Omega^{\cdot}_{X/S_0})) \Rightarrow \mathbb{H}^i(X,\Omega^{\cdot}_{X/S})$ to play a role. Amazingly, Mazur's theorem asserts that $F_{X/S}$ <u>determines</u> the (mod p) Hodge and conjugate spectral sequences, (with suitable hypotheses on X).

8.26 <u>Theorem.</u> Suppose X/S_0 is smooth and proper, where (S,I,γ) is a (torsion free) p-adic base (8.1). Let $\pi_X\colon\ H^*_{cris}(X/S) \to H^*_{DR}(X/S_0)$ be the natural map (reduction mod p), and let $\Phi\colon\ H^*_{cris}(X'/S) \to H^*_{cris}(X/S)$ be the map induced by relative Frobenius $F_{X/S}$ (8.2.1). Assume that for each $s \in S_0$, the Hodge spectral sequence of $X(s)/k(s)$ is degenerate at E_1, and that $H^*_{cris}(X/S)$ is a flat $\mathcal{O}_S$-module. Then:

(8.26.1) $\pi_{X'}$ maps $M^k = \Phi^{-1}\{p^k H^i_{cris}(X/S)\}$ onto $F^k_{Hodge}\,H^i_{DR}(X'/S_0)$.

(8.26.2) $\pi_X \circ p^{-k}$ maps $(\mathrm{Im}\ \Phi) \cap \{p^k H^i_{cris}(X/S)\}$ onto $F^{i-k}_{con}\ H^i_{DR}(X/S_0)$.

(8.26.3) The diagram below commutes:

$$\begin{array}{ccccc} M^k & \xrightarrow{\pi_{X'}} & F^k_{Hodge} H^i_{DR}(X'/S_0) & \longrightarrow & H^{i-k}(X', \Omega^k_{X'/S_0}) \\ \downarrow \Phi & & & & \downarrow C^{-1} \\ \mathrm{Im}\ \Phi \cap p^k H^i_{cris}(X/S) & \xrightarrow{\pi_X \circ p^{-k}} & F^{i-k}_{con}\ H^i_{DR}(X/S_0) & \longrightarrow & H^{i-k}(X, \mathcal{H}^k(\Omega^{\cdot}_{X/S_0})) . \end{array}$$

(The diagram makes sense because of Lemma 8.27, which tells us that the Hodge and conjugate spectral sequences degenerate suitably.)

Proof. Actually, various sets of hypotheses and conclusions are possible. For instance, (8.26.1) and (8.26.2) hold assuming only that $H^i_{cris}(X/S)$ is p-torsion free and that the conjugate spectral sequence of X/S_0 is degenerate at E_2. Thus, in the lemmas which follow, we assume only that X/S_0 is proper and smooth, and state the additional hypotheses as we need them.

If X does satisfy all the hypotheses of the theorem, the base changing theorem for crystalline cohomology and the flatness assumption show that $H^*_{cris}(X/S) \otimes_{\mathcal{O}_S} \mathcal{O}_{S_0} \cong H^*_{DR}(X/S_0)$, and in particular, the latter is locally free. Therefore, X/S_0 satisfies the hypotheses of the following lemma:

8.27 Lemma. Suppose $H^*_{DR}(X/S_0)$ is a flat $\mathcal{O}_{S_0}$-module and that the Hodge spectral sequence of $X(s)/k(s)$ is degenerate at E_1, for each $s \in S_0$. Then:

8.27.1 The Hodge spectral sequence of X/S_0 is degenerate at E_1, consists of flat $\mathcal{O}_{S_0}$-modules, and commutes with arbitrary base change $T \to S_0$.

8.27.2 The conjugate spectral sequence of X/S_0 is degenerate at E_2, consists of flat $\mathcal{O}_{S_0}$-modules, and commutes with arbitrary base change.

<u>Proof.</u> Fix an integer $k \geq 0$, and consider the following three statements:

(a_k) The map $\mathbb{H}^*(X, F^k\Omega^{\cdot}_{X/S_0}) \to F^k\,\mathbb{H}^*(X,\Omega^{\cdot}_{X/S_0})$ is an isomorphism.

(b_k) $F^k\,\mathbb{H}^*(X,\Omega^{\cdot}_{X/S_0})$ is flat.

(c_k) $\mathbb{H}^{*-k}(X,\Omega^k_{X/S_0})$ is flat.

For $k=0$, (a_k) is trivial and (b_k) is given. Let us proceed by induction on k. First observe that (a_k) + (b_k) implies (c_k). Indeed, $\mathbb{H}^i(X,F^k\Omega^{\cdot}_{X/S_0})$ is flat for all i, and hence its formation commutes with arbitrary base change $T \to S_0$. Consider the diagram:

$$\begin{array}{ccc} \mathbb{H}^*(X,F^k\Omega^{\cdot}_{X/S_0}) & \xrightarrow{\alpha} & \mathbb{H}^*((X(s)),F^k\Omega^{\cdot}_{X(s)/k(s)}) \\ \downarrow{\scriptstyle\varepsilon} & & \downarrow{\scriptstyle\varepsilon(s)} \\ \mathbb{H}^{*-k}(X,\Omega^k_{X/S_0}) & \xrightarrow{\gamma} & \mathbb{H}^{*-k}((X(s)),\Omega^k_{X(s)/k(s)}) \end{array} .$$

The base changing we just established implies that α is surjective; the degeneracy of the Hodge spectral sequence of $X(s)/k(s)$ implies that $\varepsilon(s)$ is surjective. This implies that

γ is surjective. The theorem of exchange [EGA] then implies that $\mathbb{H}^{*-k}(X,\Omega^k_{X/S_0})$ is locally free and commutes with arbitrary base change, so that we have (c_k). Let us next deduce (a_{k+1}) and (b_{k+1}). Observe that the above implies that ε induces a surjection to $\mathbb{H}^*(X,\Omega^k_{X/S_0}) \otimes k(s)$, and so by Nakayama's lemma, ε is surjective. We obtain a short exact sequence:

$$0 \longrightarrow \mathbb{H}^*(X,F^{k+1}\Omega^{\cdot}_{X/S_0}) \longrightarrow \mathbb{H}^*(X,F^k\Omega^{\cdot}_{X/S_0}) \xrightarrow{\varepsilon} \mathbb{H}^{*-k}(X,\Omega^k_{X/S_0}) \longrightarrow 0$$

(a_{k+1}) follows immediately. Assumption (a_k) says that $\mathbb{H}^*(X,F^k\Omega^{\cdot}_{X/S_0}) \cong F^k\mathbb{H}^*(X,\Omega^{\cdot}_{X/S_0})$, and (a_k) and the above sequence imply (b_{k+1}). The statements (a), (b), and (c) are thus valid for all k. Taken together, they imply (8.27.1).

Now we use the Cartier isomorphism:

$$C^{-1}: H^p(X'(s), \Omega^q_{X'(s)/k(s)}) \to H^p(X(s), \mathcal{H}^q(\Omega^{\cdot}_{X(s)/k(s)}))$$

Since $X'(s) \cong X(s)\times_{k(s)}k(s)$ (via the absolute Frobenius of $k(s)$) we see that $H^p(X'(s), \Omega^q_{X'(s)/k(s)}) \cong k(s)\otimes_{k(s)}H^p(X(s), \Omega^q_{X(s)/k(s)})$. Thus, $E_{2\,\mathrm{con}}^{pq}(X(s)/k(s))$ has the same dimension as $E_{1\,\mathrm{Hodge}}^{pq}(X(s)/k(s))$. Counting dimensions shows that the conjugate spectral sequence of each $X(s)/k(s)$ degenerates at E_2. Now this spectral sequence is, after renumbering, the spectral sequence of the canonical filtration [1,1.4] $T.$ on $\Omega^{\cdot}$. Moreover, an easy argument using the Cartier isomorphism shows that the complexes $T.$ consist of coherent and flat $\mathcal{O}_{X'}$-modules. Therefore the same argument as before can be applied to $T.$, and (8.27.2) follows. □

Theorem (8.26) will follow from the local theorem (8.20) applied to some simple gauges, and from the calculus of cogauges (8.23) ff. Let us introduce the following notation:

8.28 $\varepsilon_r(i) = \{1 \text{ if } i < r,\ 0 \text{ if } i \geq r\}$

$\eta_r(i) = \{\varepsilon_r(i) + i \text{ if } i \geq 0,\ 0 \text{ if } i \leq 0\}$

$\zeta_r(i) = \{0 \text{ if } i \leq r,\ i-r \text{ if } i > r\}$.

$$\mathbb{L}\eta = \mathbb{L}\eta\, \mathbb{R}u_{X/S*}\mathcal{O}_{X/S} \quad .$$

We have a picture:

(8.28.1)

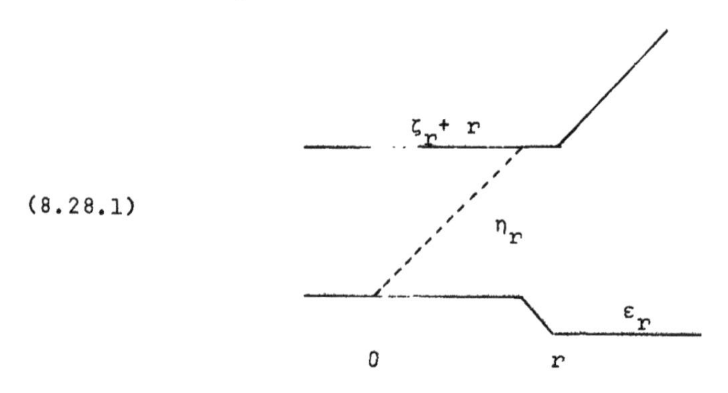

Notice that ε_{r+1} is a simple augmentation of ε_r at r . This is (essentially) the only case of a simple augmentation of gauges in which we can find a global analogue of the diagram (8.10), expressing the behaviour of Ψ_ε under simple augmentations. I find it convenient to write triangles as short exact sequences.

8.29 <u>Lemma.</u> There is a natural isomorphism of triangles:

$$\begin{array}{ccccccccc}
0 \longrightarrow & \mathbb{R}u_{X'/S*}J^{[\varepsilon_{r+1}]} & \longrightarrow & \mathbb{R}u_{X'/S}J^{[\varepsilon_r]} & \longrightarrow & \Omega^r_{X'/S_0}[-r] & \longrightarrow & 0 \\
& \Big\downarrow \Psi_{\varepsilon_{r+1}} & & \Big\downarrow \Psi_{\varepsilon_r} & & \Big\downarrow C^{-1} & & \\
0 \longrightarrow & \mathbb{L}\eta_{r+1}\mathbb{R}u_{X/S*}\mathcal{O}_{X/S} & \longrightarrow & \mathbb{L}\eta_r\mathbb{R}u_{X/S*}\mathcal{O}_{X/S} & \longrightarrow & \mathcal{H}^r(\Omega^{\cdot}_{X/S_0}[-r]) & \longrightarrow & 0
\end{array}$$

Proof. I claim that there is a canonical isomorphism:

$$J^{[\varepsilon_r]}_{X'/S}\Big/J^{[\varepsilon_{r+1}]}_{X'/S} \xrightarrow{\cong} i_{0*}J^{[r]}_{X'/S_0}\Big/J^{[r+1]}_{X'/S_0}$$

where $i_0\colon (X/S_0)_{cris} \to (X/S)_{cris}$ is the natural map. Indeed, recall that if $T \in \mathrm{Cris}(X/S)$, $i_0^*(T)$ is represented by $T_0 = T\times_S S_0$. Now $J_T^{[\varepsilon_r]} = p\mathcal{O}_T + J_T^{[r]}$, so $J_T^{[\varepsilon_r]}\Big/J_T^{[\varepsilon_{r+1}]} \cong J_T^{[r]}\Big/(p\mathcal{O}_T) \cap J_T^{[r]} + J_T^{[r+1]} \cong J_{T_0}^{[r]}\Big/J_{T_0}^{[r+1]}$. Since $(i_{0*}F)_T = F_{T_0}$ for any abelian F, this proves the claim. Clearly i_{0*} is exact. Thus, we have canonically:

$$\mathbb{R}u_{X'/S*}(J^{[\varepsilon_r]}_{X'/S}\Big/J^{[\varepsilon_{r+1}]}_{X'/S}) \cong \mathbb{R}u_{X/S*}\mathbb{R}i_{0*}J^{[r]}_{X'/S_0}\Big/J^{[r+1]}_{X'/S_0}$$
$$\cong \mathbb{R}u_{X'/S_{0*}}J^{[r]}_{X'/S_0}\Big/J^{[r+1]}_{X'/S_0} \cong \Omega^r_{X'/S_0}[-r] .$$

Plugging this into the proper place in the triangle induced by applying $\mathbb{R}u_{X'/S*}$ to the exact sequence:

$$0 \to J^{[\varepsilon_{r+1}]}_{X'/S} \to J^{[\varepsilon_r]}_{X/S} \to J^{[\varepsilon_r]}_{X'/S}\Big/J^{[\varepsilon_{r+1}]}_{X'/S} \to 0 ,$$

you get the top triangle of the lemma. The bottom one follows from (8.23.6). To check that the diagram commutes, we are reduced to the local calculation (8.10). □

8.30 <u>Lemma.</u> Suppose that $H^*_{cris}(X'/S)$ is p-torsion free.

8.30.1 The map $\pi'_r : H^*(X'/S, J^{[\varepsilon_r]}_{X'/S}) \to \mathbb{H}^*(X', F^r\Omega^{\cdot}_{X'/S_0})$ is surjective.

8.30.2 The image $H^*_{\varepsilon_r}$ of $H^*(X'/S, J^{[\varepsilon_r]}_{X'/S})$ in $H^*_{cris}(X'/S)$ is the inverse image of $F^rH^*_{DR}(X'/S_0)$.

8.30.3 If, additionally, the Hodge spectral sequence of X'/S_0 degenerates at E_1, the map α_r: $H^*(X'/S, J^{[\varepsilon_r]}_{X'/S}) \to H^*(X'/S, \mathcal{O}_{X'/S})$ is injective.

Proof. For $r = 0$, the first statement is a consequence of the base changing formula, and the other statements are trivial. There is an exact ladder:

$$\begin{array}{ccccccc}
H^{i-r-1}(X', \Omega^r_{X'/S_0}) & \longrightarrow & H^i(X'/S, J^{[\varepsilon_{r+1}]}_{X'/S}) & \to & H^i(X'/S, J^{[\varepsilon_r]}_{X'/S}) & \longrightarrow & H^{i-r}(X', \Omega^r_{X'/S_0}) \\
\downarrow \cong & & \downarrow \pi'_{r+1} & & \downarrow \pi'_r & & \downarrow \cong \\
H^{i-r-1}(X', \Omega^r_{X'/S_0}) & \longrightarrow & H^i(X', F^{r+1}\Omega^{\cdot}_{X'/S_0}) & \to & H^i(X', F^r\Omega^{\cdot}_{X'/S_0}) & \to & H^{i-r}(X', \Omega^r_{X'/S_0})
\end{array}$$

This diagram proves (8.30.1), by induction on r. This immediately implies (8.30.2) because $p\mathcal{O}_{X'/S} \subseteq J^{[\varepsilon_r]}_{X'/S}$, hence $pH^*_{cris}(X'/S) \subseteq H^*_{\varepsilon_r}$. If the Hodge spectral sequence of X'/S_0 degenerates at E_1, one sees from the diagram that the maps $H^i(X'/S, J^{[\varepsilon_r]}_{X'/S}) \to H^{i-r}(X', \Omega^r_{X'/S_0})$ are all surjective, hence (8.30.3) follows by induction. □

Thanks to the local theorem (8.20), we have an enormous diagram:

(8.31)

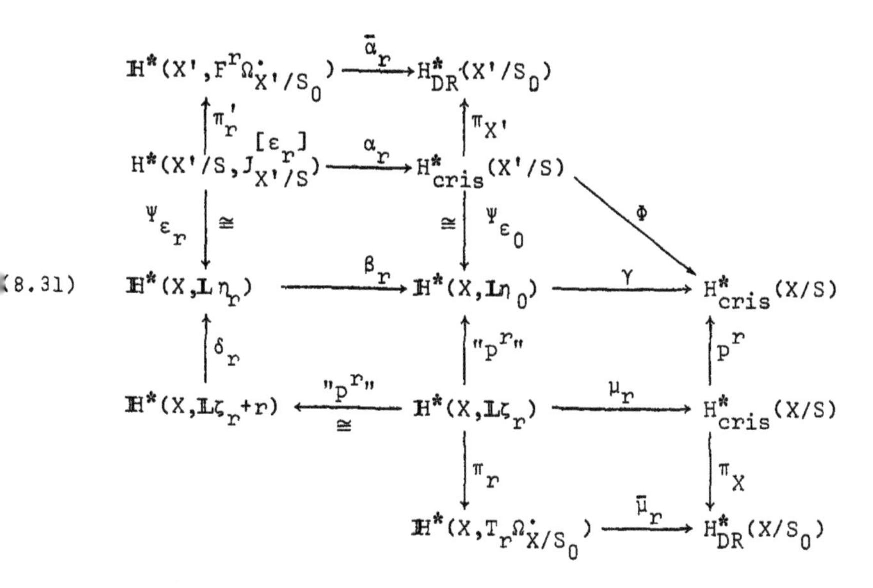

Consider the hypotheses:

(a) $H^*_{cris}(X/S)$ is p-torsion free.

(b) The conjugate spectral sequence of X/S_0 is degenerate at E_2.

(c) The Hodge spectral sequence of X'/S_0 is degenerate at E_1.

8.32 Lemma. Hypotheses (a) and (b) imply (8.26.1) and (8.26.2) of the theorem.

Proof. Proposition (8.24) tells us that if η is any co-gauge, the map $\mathbb{H}^*(X, \mathbb{L}\eta) \to H^*_{cris}(X/S)$ is an isomorphism onto its image H_η, and that $\eta \mapsto H_\eta$ is compatible with the lattice operations.

To prove (8.26.1), let $H^*_{\varepsilon_r}$ denote the image of α_r, which by Lemma (8.30) is the inverse image of $F^r_{Hodge} H^*_{DR}(X'/S_0)$. We have to prove that the H_{ε_r} and M^r have the same image in $H_{DR}(X'/S_0)$, i.e., that $H_{\varepsilon_r} = M^r + pH^*_{cris}(X'/S)$. Contemplate diagram (8.31). Since Ψ_{ε_0} and Ψ_{ε_r} are isomorphisms and γ is injective, this is equivalent to: $H_{\eta_r} = \Phi(M^r) + p\, H_{\eta_0}$. But $\Phi(M^r) + p\, H_{\eta_0} = p^r H_{cris} \cap \mathrm{Im}(\Phi) + p\, H_{\eta_0} = H_r \cap H_{\eta_0} + p\, H_{\eta_0} = H_{r \vee \eta_0} + H_{\eta_0 + 1} = H_{(\zeta_r + r) \wedge \eta_0 + 1} = H_{\eta_r}$. (Notice: $\Phi(M^r) = H_{(\zeta_r + r)}$.)

To prove (8.26.2), first note that $\mathrm{Im}(\Phi) \cap p^r H_{cris}(X/S) = H_{\eta_0} \cap H_r = H_{\zeta_r + r}$, and $p^{-r} H_{\zeta_r + r} \cong H_{\zeta_r}$. So it suffices to prove that π_r maps H_{ζ_r} onto $\mathbb{H}^*(X, T_r \Omega^{\cdot}_{X/S_0})$. (Recall that the image of $\mathbb{H}^i(X, T_r)$ in $H^i_{DR}(X/S_0)$ is $F^{i-r}_{con} H^i_{DR}(X/S_0)$.) But (8.23.5) gives us an exact sequence:

$$\cdots \to H^i(X, \mathbb{L}\zeta_r \vee 1) \xrightarrow{\rho} H^i(X, \mathbb{L}\zeta_r) \xrightarrow{\pi_r} H^i(X, T_r) \to \cdots .$$

Since Proposition (8.24) implies that the ρ's are injective, the π_r's are surjective. □

8.33 Remark. Assuming only that the conjugate spectral sequence of X/S_0 is degenerate at E_2, it is still possible to prove part of (8.26.1) and (8.26.2):

8.33.1 $\pi_{X'}(M^r) \subseteq F^r_{Hodge} H^*_{DR}(X'/S_0)$.

8.33.2 If $x \in \mathrm{Im}(\Phi) \cap p^r H^i_{cris}(X/S)$, then $x = p^r y$ for some y such that $\pi_X(y) \in F^{i-r}_{con} H^i_{DR}(X/S_0)$.

Since this remark seems useless, we do not include its proof.

The following lemma completes the proof of Theorem (8.26).

8.34 <u>Lemma.</u> Hypotheses (a), (b), and (c) imply (8.26.3).

<u>Proof.</u> We have a commutative diagram of triangles:

$$\begin{array}{ccccccccc}
0 & \longrightarrow & \mathbb{L}\zeta_r \vee 1 & \longrightarrow & \mathbb{L}\zeta_r & \longrightarrow & T_r & \longrightarrow & 0 \\
 & & \wr\| \; p^r & & \wr\| \; p^r & & \| & & \\
0 & \longrightarrow & \mathbb{L}(\zeta_r \vee 1 + r) & \longrightarrow & \mathbb{L}(\zeta_r + r) & \longrightarrow & T_r & \longrightarrow & 0 \\
 & & \downarrow & & \downarrow & & \downarrow & & \\
0 & \longrightarrow & \mathbb{L}\eta_{r+1} & \longrightarrow & \mathbb{L}\eta_r & \longrightarrow & \mathcal{H}^r & \longrightarrow & 0 \\
 & & \uparrow\cong & & \uparrow\cong & & \cong\uparrow C^{-1} & & \\
0 & \longrightarrow & \mathbb{R}u_*J^{[\varepsilon_{r+1}]} & \longrightarrow & \mathbb{R}u_*J^{[\varepsilon_r]} & \longrightarrow & \Omega^r & \longrightarrow & 0
\end{array}$$

Moreover, M^r is contained in the image of the injective map α_r, so that $M^r \hookrightarrow H^*_{cris}(X/S)$ factors through α_r. Recalling from (8.32) that $\Phi(M_r) = H_{(\zeta_r+r)}$, we obtain a commutative:

$$\begin{array}{ccccc}
M^r & \longrightarrow & H(J^{[\varepsilon_r]}) & \longrightarrow & H(\Omega^r) \\
\downarrow & & \downarrow & & \downarrow C^{-1} \\
H(\zeta_r + r) & \longrightarrow & H(\eta_r) & \longrightarrow & H(\mathcal{H}^r) \\
p^r \uparrow\cong & & & \nearrow & \\
H(\zeta_r) & \longrightarrow & H(T_r) & &
\end{array}$$

□

It is clear that Theorem (8.26) implies that the relative Frobenius morphism $\Phi\colon H^*_{cris}(X'/S) \to H^*_{cris}(X/S)$ determines the (mod p) Hodge filtration of X'/S_0 and conjugate filtration of X/S_0, assuming the stated degeneracy and torsion hypotheses. Even without these hypotheses we can find a relationship, in the form of inequalities, between the p-adic divisibility properties of Φ and the Hodge numbers of X.

For simplicity, we restrict attention to the case in which S is the formal spectrum of the p-Witt ring W of a perfect field k, and X is smooth and proper over $S_0 = \mathit{Spec}\ k$. Then the absolute Frobenius endomorphism F_X of X induces a σ-linear endomorphism T of $H^*_{cris}(X/S)$, where σ is the Frobenius automorphism of W. Moreover, it is easy to see that $H^*_{cris}(X'/S) = H^*_{cris}(X/S) \underset{\sigma}{\otimes} W$, and Φ is the obvious linearization of T. It is a well known consequence of Poincaré duality that Φ (and hence F) is injective modulo torsion — and we shall indicate another proof of this fact below. Killing torsion, we thus ket what Mazur calls and F-crystal, and its associated span.

8.35 Definition. An "F-crystal over W" is an injective σ-linear endomorphism T of a free finite rank W-module M. A "span over W" is an injective W-linear map of free finite rank W-modules of the same rank. If $T\colon M \to M$ is an F-crystal, its associated F-span is its linearization $M \underset{\sigma}{\otimes} W \to W$.

The span of an F-crystal measures its p-adic divisibility. Precisely, if $M' \hookrightarrow M$ is any span, there is a direct sum decomposition $M = \bigoplus_{i=0}^{\infty} M^i$, such that $\mathrm{Im}(M') = \bigoplus_{i=0}^{\infty} p^i M^i$. The ranks e^i of M^i then determine the span up to isomorphism. They are called the "Hodge numbers" of the span.

In order to state the inequalities between the Hodge numbers of the crystalline span:

$$\Phi: H^*_{cris}(X'/S)/(\text{torsion}) \longrightarrow H^*_{cris}(X/S)/(\text{torsion})$$

and the Hodge numbers of X/k, it is convenient to introduce the so-called "Hodge polygon" defined by a sequence $(a_0, a_1, \ldots)$ of nonnegative integers. This is the continuous graph consisting of the straight line segment of slope 0 over the interval $[0, a_0]$, of slope 1 over the interval $[a_0 + a_1] \ldots$, and slope i over $\left[\sum_{j=0}^{i-1} a_j, \sum_{j=0}^{i} a_j\right]$. The "Hodge polygon of X/K" (in degree n) means the Hodge polygon determined by the numbers $h^q = \dim_k H^{n-q}(X, \Omega^q_{X/k})$; the "Hodge polygon of $(H^n_{cris}(X/S), F_X)$" means the Hodge polygon determined by the Hodge numbers e^q of its associated span.

8.36 Theorem. If X/k is smooth and proper, the Hodge polygon of $(H^n_{cris}(X/S), F_X)$ lies on or above the Hodge polygon of X/k. (In particular, the endpoint of the first lies to the left of the endpoint of the second.)

Proof. To prove the theorem, and perhaps to give some insight into its meaning, it is helpful to baldly list the inequalities which it asserts.

8.37 Lemma. If $(a_0, a_1 \ldots)$ and $(b_0, b_1 \ldots)$ are sequences of nonnegative integers defining respective Hodge polygons A and B, then B lies on or above A iff for all $m \geq 1$:

$$I(m): \quad ma_0 + (m-1)a_1 + \cdots + a_{m-1} \geq mb_0 + (m-1)b_1 + \cdots + b_{m-1} \quad .$$

Proof. Suppose the inequalities $I(m)$ hold for all m. Observe first that the domain of B is contained in the domain of A, i.e. $a_0 + a_1 + \cdots \geq b_0 + b_1 + \cdots$. Indeed, this is trivial unless there exists a k such that $a_j = 0$ for $j > k$, and in that case $I(m)$ implies that whenever $m > \max\{k,n\}$ $ma_0 + (m-1)a_1 + \cdots + (m-k)a_k \geq mb_0 + (m-1)b_1 + \cdots (m-n)b_n$. Taking the limit as m approaches infinity, we see that $a_0 + a_1 + \cdots + a_k \geq b_0 + b_1 + \cdots b_n$ for any n, as required.

Now to prove that B lies over A, it suffices to check that each of the endpoints of the line segments comprising B lie above A, since the region above A is convex. Suppose $(i, B(i))$ is such an endpoint, so $i = b_0 + b_1 + \cdots + b_n$ for some n, and $B(i) = b_1 + 2b_2 + \cdots + nb_n$. To calculate $A(i)$, note from the previous paragraph that there exists an m with $a_0 + a_1 + \cdots a_{m-1} \leq i \leq a_0 + a_1 + \cdots + a_m$, and then $A(i) = a_1 + 2a_2 + \cdots + (m-1)a_m + m[i - (a_0 + \cdots + a_{m-1}]$. Thus the desired inequality $A(i) \leq B(i)$ reduces to $mb_0 + (m-1)b_1 + \cdots (m-n)b_n \leq ma_0 + \cdots a_{m-1}$, which is an immediate consequence of $I(m)$. The proof of the converse is similar — investigate the consequences of the fact that the breakpoints of A lie under B. □

8.37.1 Remark. It is perhaps more natural to work with the partial sums $s_m = a_0 + \cdots + a_m$, for all m . Then if $t_m = b_0 + \cdots + b_m$, one sees from the above that $A \leq B$ if $s_0 + \cdots + s_m \geq t_0 + \cdots + t_m$ for all m .

8.37.2 If $M' \longrightarrow M$ is a span with Hodge numbers $e^0, e^1 \ldots$ it is trivial to observe that

$$\text{length}(M'/p^n M \cap M' = me^0 + (m-1)e^1 + \cdots + e^{m-1} .$$

It is now clear that the following lemma proves Theorem (8.36). Note that it also implies that Φ really is injective.

3.38 Lemma. Let $M = H^i_{cris}(X/W)/(\text{torsion})$, and let M' be the image in M of Φ. Then M/M' has finite length. Moreover, for any m, $\text{length}(M'/p^m M \cap M') \leq mh^0 + (m-1)h^1 + \cdots + h^{m-1}$.

Proof. We leave the first statement as an exercise for the reader – use (8.28) and the method of the following proof of the second statement.

Use the notation (8.28), and notice that $\eta_0 \vee m = \zeta_m + m$. We have a commutative diagram:

$$\begin{array}{ccccc} \mathbb{H}^i(X, \mathbb{L}\zeta_m + m) & \longrightarrow & \mathbb{H}^i(X, \mathbb{L}\eta_0) & \xrightarrow{q} & \mathbb{H}^i(X, \mathbb{L}\eta_0/\zeta_m + m) \\ \downarrow & & \downarrow \gamma & & \\ \mathbb{H}^i(X, \mathbb{L}\, m) & \longrightarrow & H^i_{cris}(X/W) & & \end{array}$$

Theorem (8.20) implies that the image of γ in M is exactly M', and the diagram implies that the image of $\mathbb{H}^i(X, \mathbb{L}\zeta_m + m)$ in M' is contained in $p^m M \cap M'$. Thus, we get an induced map from the image of q to $M'/p^m M \cap M'$, necessarily surjective. Therefore it suffices to prove that $\mathbb{H}^i(X, \mathbb{L}\eta_0/\zeta_m + m)$ has length $\leq mh^0 + \cdots + h^{m-1}$.

If $m = 0$, $\eta_0 = \zeta_0$, so the statement is trivial. We proceed by induction on m, using the exact sequence:

$$\mathbb{H}^i(X, \mathbb{L}\zeta_{m-1}+m-1/\zeta_m+m) \to \mathbb{H}^i(X, \mathbb{L}\eta_0/\zeta_m+m) \to \mathbb{H}^i(X, \mathbb{L}\eta_0/\mathbb{L}\zeta_{m-1}+m-1).$$

But $\mathbb{H}^i(X, \mathbb{L}\zeta_{m-1}+m-1/\zeta_m+m) \cong \mathbb{H}^i(X, \mathbb{L}\zeta_{m-1}/\zeta_m+1) \cong \mathbb{H}^i(X, T_{m-1}\Omega^{\cdot}_{X/k})$ by (8.23.5). Since the length of this is less than or equal to $h^0 + h^1 + \cdots + h^{m-1}$ (by the Cartier isomorphism), the desired inequality follows immediately from the induction assumption. The reader may wish to note that we have in fact proved a slightly stronger inequality than claimed, which he can work out according to his needs. □

Katz's conjecture is an immediate consequence of the above result. Let me remark that this conjecture (inspired by Dwork's fundamental work) is what led Mazur to his investigations. To express it, recall that the Newton polygon of an F-crystal $T: M \to M$ over W is defined as follows: Choose a basis for M, and express T as a matrix, as if it were linear. The eigenvalues of the matrix depend on the choice of basis, but their p-adic ordinals do not. For each $r \in \mathbf{Q}$, let c^r be the number of eigenvalues with ordinal r, and form the continuous graph which begins at (0,0), with slope r over an interval of length c_r, arranged in increasing order of r. This (convex) graph is called the Newton polygon of the F-crystal $T: M \to M$. It is worth remarking that it depends, in fact, only on $T \otimes id_\Omega: M \otimes \Omega \to M \otimes \Omega$, where Ω is the fraction field of $W(\bar{k})$ In fact a theorem of Dieudonné and Manin [3] asserts that the Newton polygon of $T \otimes id_\Omega$ classifies it up to isomorphism. Katz conjectured the following:

8.39 Theorem. If X/k is smooth and proper, and $M = H^*_{cris}(X/W)/(\text{torsion})$, then the Newton polygon of (M, F^*_X) lies on or above the Hodge polygon of X/k.

Proof. Thanks to the previous result, we can follow Mazur and reduce to an elementary calculation in linear algebra:

8.40 Lemma. The Newton polygon of an F-crystal lies on or above its Hodge polygon, and both have the same endpoint.

Proof. There is another way to calculate the Newton polygon of an F-crystal which is often more convenient: Choose a matrix representative of T as above, and let $p(X) = X^n + a_1X^{n-1} + \cdots + a_n$ be the characteristic polynomial of the matrix. Plot the points $(i, \text{ord}_p(a_i))$; then the Newton polygon of T is the convex hull of this graph. (We leave this verification for the reader.)

Now suppose that the span $M' \hookrightarrow M$ of T has Hodge numbers $e^0, e^1, \ldots$. It is clear that the span of $\Lambda^i T$ is $\Lambda^i M' \longrightarrow \Lambda^i M$, and it is easy to calculate the Hodge numbers of this span. In particular, the index of its first nonzero Hodge number is clearly

$$e^1 + 2e^2 + \cdots + (j-1)e^{j-1} + j[i - (e^0 + \cdots + e^{j-1})] ,$$

if $e^0 + \cdots + e^{j-1} \leq i \leq e^0 + \cdots + e^j$. Note that this is just the value $E(i)$ of the Hodge polygon of $M' \longrightarrow M$ at i . In particular, $\Lambda^i T$ is divisible by $p^{E(i)}$, and hence trace $\Lambda^i T = (-1)^i a_i$ has $\text{ord} \geq E(i)$. Since the Hodge polygon is convex, and since $\{(i, \text{ord}_p(a_i))\}$ lies over it, so does the convex hull — i.e. the Newton polygon. By looking at the highest exterior power, one gets the statement about the endpoints. □

(Added December, 1977). It seems to be worthwhile to prove a stronger version of (8.26), which tells us that even over a p-adic base, the filtration of crystalline cohomology provided by $\mathcal{J}_{X/S}$ is determined by the action of Frobenius. The reader who so desires can also prove the intermediate statements (mod p^n for all n).

(8.41) Theorem: With the hypotheses and notation of (8.26):

(8.41.1) For any tame gauge ε, the map

$$H^i_{cris}(X'/S, \mathcal{J}^{[\varepsilon]}_{X'/S}) \to H^i_{cris}(X'/S, \mathcal{O}_{X'/S}) \text{ is injective,}$$

and its image is $M^{\varepsilon} H^i_{cris}(X'/S)$.

(8.41.2) For any $k \geq 0$, we have:

$$H^i_{cris}(X'/S, \mathcal{J}^{[k]}_{X'/S}) \xrightarrow{\sim} \sum_{j=0}^{\infty} p^{<k-j>} M^j. \text{ In particular}$$

$$H^i_{cris}(X'/S, \mathcal{J}^{[k]}_{X'/S}) = M^k, \text{ if } k < p.$$

Proof: Let us first remark that (8.41.2) is the special case of (8.41.1) obtained by setting ε = the maximal tame gauge with cut-off at i (8.18.3). For the second part of (8.41.2), simply note that $p^r M^j \subseteq M^{j+r}$.

To prove (8.41.1), let $\eta = \varepsilon + id$, and consider the diagram:

$$\begin{array}{ccccc} H^i(X'/S, \mathcal{J}^{[\varepsilon]}_{X'/S}) & \to & H^i(X'/S, \mathcal{O}_{X'/S}) & & \Phi \\ H^i(\psi_\varepsilon) \downarrow \cong & & H^i(\psi_o) \downarrow \cong & \searrow & \\ H^i(X/S, \mathbb{L}\eta\, \mathbb{R}u_{X/S*}\mathcal{O}_{X/S}) & \to & H^i(X/S, \mathbb{L}id\ \mathbb{R}u_{X/S*}\mathcal{O}_{X/S}) & \to & H^i(X/S, \mathbb{R}u_{X/S*}\mathcal{O}_{X/S}) \end{array}$$

Theorem (8.20) tells us that the vertical arrows are isomorphisms, and Proposition (8.24) tells us that the horizontal arrows along the bottom are injective. This implies the injectivity of the horizontal arrows along the top, as well as the injectivity of Φ .

To identify the image of $H^i(X'/S, \mathcal{J}^{[\epsilon]}_{X'/S})$ as M^{ϵ}, it clearly suffices to prove that $H_{\eta} = \Phi\,(M^{\epsilon})$, in the notation of (8.24). Now:

$$\Phi\,(M^{r}) = p^{r} \cap \operatorname{Im}(\Phi) = p^{r} \cap H_{id} = H_{id \vee r}\text{, by (8.24).}$$

Hence: $\Phi\,(M^{\epsilon}) = \sum_{r} p^{\epsilon(r)}\,\Phi\,(M^{r}) = H_{\beta}$, where $\beta = \inf\{\beta_r\}$ and $\beta_r = (id \vee r) + \epsilon\,(r)$. But if $r \leq i$, $\beta_r\,(i) = i + \epsilon(r) \geq r + \epsilon\,(r) = \eta\,(r)$, with equality if $r=i$, and if $r \geq i$, $\beta_r(i) = r + \epsilon\,(r) = \eta\,(r) \geq \eta\,(i)$. Thus $\beta = \eta$, and the theorem is proved. ///

References for Chapter 8

[1] Deligne, P. "Théorie de Hodge II" Publ. Math. I.H.E.S. 40, (1971), 5-58.

[2] Katz, N. "Nilpotent Connections and the Monodromy Theorem....." Publ. Math. I.H.E.S 39 (1970) 175-232.

[3] Manin, Yu.: "The theory of commutative formal groups over fields of finite characteristic" Russian Math. Surv. 18 (1963).

[4] Mazur, B. "Frobenius and the Hodge Filtration" Bull. A.M.S. 78 (1972) 653-667.

[5] ________ "Frobenius and the Hodge Filtration – estimates" Ann. of Math. 98 (1973) 58-95.

Appendix A

The Construction of $\Gamma_A(M)$

This appendix is devoted to a brief exposition of two papers of N. Roby [1,2] in which the many interesting properties of the functor $\Gamma_A(M)$ are developed.

Let M be an A-module. Form $G = G_A(M)$, the polynomial A-algebra on the set of indeterminates $\{(x,n): x \in M,\ n \in \mathbb{N}\}$. Let $I = I_A(M)$ be the ideal in G generated by the following four types of elements:

I. $(x,0) - 1$.

II. $(\lambda x,n) - \lambda^n(x,n)$

III. $(x,n)(x,m) - ((n,m))(x,n+m)$, where $((n,m)) = \frac{(n+m)!}{n!m!}$

IV. $(x+y,n) - \sum_{i+j=n} (x,i)(y,j)$.

Let $\Gamma_A(M) = G_A(M)/I_A(M)$, and for each $x \in M$, let $x^{[n]}$ denote the image of (x,n) in $\Gamma_A(M)$. If we assign to (x,n) degree n, the algebra $G_A(M)$ becomes graded, and the ideal $I_A(M)$ a homogeneous ideal. Hence $\Gamma = \Gamma_A(M)$ is also a graded A-algebra, and $x^{[n]}$ has degree n. It is easy to see that $\Gamma_0(M) \cong A$ and that $\Gamma_1(M) \cong M$.

We want to show that there is a PD structure γ on the ideal $\Gamma^+(M) = \bigoplus_{i>0} \Gamma_i(M)$, with $\gamma_n(x^{[1]}) = x^{[n]}$. Before doing so, it is convenient to investigate three interesting mapping properties of the functor Γ_A. The first of these shows that Γ_A is, in a way, a multiplicative version of Sym_A .

Definition. *If R is an A-algebra and $f \in R[[T]]$, we say* that f is "of exponential type" iff $f(0) = 1$ and $f(T_1 + T_2) = f(T_1)f(T_2)$, for indeterminates T_1 and T_2. The set of all such f is denoted by "exp R". It is a subgroup of the group of units of $R[[T]]$, and an R-module using the rule $(rf)(T) = f(rT)$.

We can think of exp(R) as the set of one-parameter subgroups of $\hat{\mathbb{G}}_m(R)$. The first important universal characterization of $\Gamma_A(R)$ is the following:

(A1) Proposition. There is a natural bijection between $\{$A-algebra maps: $\Gamma_A(M) \xrightarrow{\alpha} R\}$ and $\{$A-module maps: $M \xrightarrow{\beta} \exp(R)\}$. This correspondence is given as follows: If $x \in M$,
$$\beta(x) = \sum_{x=0}^{\infty} \alpha(x^{[n]})T^n .$$

Proof. A map of sets $\beta : M \to R[[T]]$ assigns to each $x \in M$ an element $\beta(x) = \sum_{n=0}^{\infty} b(x,n)T^n$ of $R[[T]]$, and we can extend b to a unique A-algebra map $b : G_A(M) \to R$. Indeed, the set of all such β is thus naturally isomorphic to the set of all A-algebra maps $G_A(M) \to R$; what needs to be shown is that β comes from an A-linear map $M \to \exp(R)$ iff b kills $I_A(M)$. The map β will be additive iff $\beta(x+y) = \beta(x)\beta(y)$, i.e. iff $\sum_{i+j=n} b(x,i)b(y,j) = b(x+y,n)$, i.e. iff b kills the generators of type IV. A-linearity is, similarly, equivalent to annihilation of the generators of type II. Finally, $\beta(x)$ is of exponential type if $\beta(x)(0) = 1$ (type I) and

$\beta(x)(T_1)\beta(x)(T_2) = \beta(x)(T_1 + T_2)$, which says that:

$$\sum_{i,j} b(x,i)b(x,j)T_1^i T_2^j = \sum_n b(x,n)(T_1+T_2)^n = \sum_n b(x,n) \sum_{i+j=n} ((i,j))T_1^i T_2^j$$

i.e. type III $\subseteq$ Ker (b). □

Immediate corollaries are the following important compatibilities:

(A2) Proposition.

1) If A' is an A-algebra, $A' \otimes_A \Gamma_A(M) \cong \Gamma_{A'}(A' \otimes_A M)$.

2) If $\{N_\lambda : \lambda \in \Lambda\}$ is a direct system of A-modules,

$$\varinjlim \Gamma_A(M_\lambda) \cong \Gamma_A(\varinjlim M_\lambda)$$

3) $\Gamma_A(M) \otimes \Gamma_A(N) \cong \Gamma_A(M \oplus N)$. □

(A3) Proposition. Suppose M is free with basis $S = \{x_i : i \in I\}$. Then $\Gamma_n(M)$ is free with basis $\{x_1^{[q_1]} x_2^{[q_2]} \ldots x^{[q_k]} : \Sigma q_i = n\}$.

Proof. Using 2) above we reduce to the case in which S is finite, using 3) we reduce to the case in which S has cardinality 1, and using 1) we reduce to the case $A = \mathbb{Z}$. It is clear that if $S = \{x\}$, $x^{[n]}$ generates $\Gamma_n(M)$, so we have only to show that $\Gamma_n(M)$ is torsion free. Since $\exp(T) = 1 + T/1 + T^2/2! + \cdots \in \exp(\mathbb{Q})$, there is a $\mathbb{Z}$-module map $M \to \exp(\mathbb{Q})$ sending x to $\exp(T)$, and hence a $\mathbb{Z}$-algebra map $\alpha : \Gamma_{\mathbb{Z}}(M) \to \mathbb{Q}$ sending $mx^{[n]}$ to $m/n! \in \mathbb{Q}$. It follows that $mx^{[n]}$ is nonzero unless $m = 0$. □

To describe the next universal mapping property of $\Gamma_A(M)$ we need Roby's notion of a "polynomial function" between two

A-modules. (There are several excellent reasons for calling them "modular functions," but I shall resist the temptation to do so.) Here is a fancy description: If M is an A-module, let $\underline{M}$ denote the corresponding quasi-coherent sheaf on the "big" Zariski topology $(\mathrm{Spec}\ A)_{ZAR}$ of all A-schemes. Then a polynomial function $M \to N$ is just a morphism of the underlying sheaves of <u>sets</u> $\underline{M} \to \underline{N}$. This boils down to the following:

(A4) <u>Definition.</u> If M and N are A-modules, a "polynomial function $f: M \to N$" is a compatible collection of set maps $\{f_R : N\otimes_A R \longrightarrow N\otimes_A R\}$ for all A-algebras R. The set of all such is denoted by $P(M,N)$. A polynomial function f is said to have "weight n" iff for all $r \in R$ and all $z \in M\otimes_A R$, $f_R(rz) = r^n f(z)$, in which case we write $f \in P_n(M,N)$

Any $f \in P(M,N)$ can be expressed as a locally finite sum of homogeneous ones. To see this, form $f_{R[T]} : M\otimes_A R[T] \to N\otimes_A R[T]$, and if $x \in M\otimes R$, write $f_{R[T]}(x\otimes T) = \sum_n f_R^n(x)T^n$, with $f_R^n(x) \in N\otimes_A R$. (In the sum, only finitely many terms are nonzero, for each x .) One sees easily that each $\{f_R^n\}$ thus defined is an element of $P_n(M,N)$, and by using the compatibility of $\{f_R\}$ with the map $R[T] \to R$ sending T to 1, one sees that $f_R(x) = \sum_n f_R^n(x)$.

We define an A-module structure on $P(M,N)$ via the structure of N. Thus, if $\alpha \in A$, $(\alpha f+g)_R(x) = \alpha f_R(x) + g_R(x)$. The notion of composition $P(M,N) \times P(N,Q) \to P(M,Q)$ is, I think, clear. In particular, linear maps define polynomial functions of weight one, and if $h: N \to N'$ is linear and $f \in P_n(M,N)$, then $h \circ f \in P_n(M,N')$.

Using the fact that formation of Γ_n commutes with base change, we see that $x \longrightarrow x^{[n]}$ is a polynomial function of weight n. More precisely, for each A-algebra R, let $\ell_{n,R}: M \otimes R \longrightarrow \Gamma_n(M) \otimes R$ be the map taking $x \in M \otimes R$ to the image of $x^{[n]}$ in $\Gamma_n(M) \otimes R$ via the inverse of the base change isomorphism: $\Gamma_n(M) \otimes R \to \Gamma_n(M \otimes R)$. It is clear that $\ell_n = \{\ell_{n,R}\}$ is an element of $P_n(M, \Gamma_n(M))$. In fact, it turns out to be the universal polynomial function of weight n. More precisely:

(A5) Theorem. There is a natural bijection:

$$\mathrm{Hom}_A[\Gamma_n(M), N] \xleftrightarrow{\circ \ell_n} P_n(M,N) .$$

In other words, if $p \in P_n(M,N)$, there is a unique A-linear $f: \Gamma_n(M) \to N$ such that for all $x \in M \otimes R$, $f_R(x^{[n]}) = p_R(x)$, compatibly with base change.

Before proving this theorem, we have to briefly study derivatives of polynomial functions. It is most convenient to deal with the entire Taylor expansion simultaneously.

If $x \in M$, if $p \in P(M,N)$, and if $z \in M \otimes R$, $p_{R[T]}(x \otimes T + z)$ is an element of $N \otimes R[T]$, which we denote by $S_x(p)_R(z)$ or $S_x[T](p)_R(z)$, if necessary. This operation is compatible with change of R, so that we can regard $S_x(p)$ as an element of $P(M, N \otimes A[T])$. Expanding in powers of T, we see that there are unique elements $D_x^{[i]}(p)_R(z)$ of $N \otimes R$ such that

$S_x(p)_R(z) = \sum_{i=0}^{\infty} D_x^{[i]}(p)_R(z)T^i$. Thus each $D_x^{[i]}(p)$ is an element of $P(M,N)$. We regard $D_x^{[i]}$ as an endomorphism of $P(M,N)$ and $S_x = \sum_{i=0}^{\infty} D_x^{[i]}T^i$ as a formal power series with coefficients in the ring of all such endomorphisms.

It is clear that $S_x(p_1+p_2) = S_x(p_1) + S_x(p_2)$, and hence that the same holds for each $D_x^{[i]}$. In fact, the operators $D_x^{[i]}$ are even A-linear, because:

$$S_x(ap)_R(z) = (ap_{R[T]})(x\otimes T+z) = a[p_{R[T]}(x\otimes T+z)] = aS_x(p)_R(z) .$$

Moreover, it is even true that the $D_x^{[i]}$'s commute with each other. If x_1 and x_2 are elements of M, and if T_1 and T_2 are indeterminates, we have for any $z \in M\otimes R$:

$$p_{R[T_1,T_2]}(x_1\otimes T_1+x_2\otimes T_2+z) = p_{R[T_2,T_1]}(x_2\otimes T_2+x_1\otimes T_1+z)$$

$$S_{x_1}(p)_{R[T_2]}(x_2\otimes T_2+z) = S_{x_2}(p)_{R[T_1]}(x_1\otimes T_1+z)$$

$$[S_{x_2}(S_{x_1}(p))]_R(z) = [S_{x_1}(S_{x_2}(p))]_R(z) . \quad \text{Hence:}$$

$$D_{x_2}^{[i]}D_{x_1}^{[j]}\, T_2^iT_1^j = D_{x_1}^{[j]}D_{x_2}^{[i]}\, T_1^jT_2^i \quad .$$

The claim follows. Note that if we take $T_1 = T_2 = T$ in the above, we can write $p_{R[T]}(x_1\otimes T+x_2\otimes T+z)$ as $S_{x_1+x_2}(p)_R(z)$, and hence we get the formula:

$$S_{x_1+x_2} = S_{x_1}S_{x_2} = S_{x_2}S_{x_1} \quad .$$

We view the latter as being the product of formal power series with coefficients in the (commutative) A-subalgebra $\mathfrak{D}$ of End(P(M,N)) generated by $\{D_x^{[i]}\}$

Lemma. S is an A-linear map $N \to \exp(\mathcal{O})$.

Proof. Recall that we are regarding S_x as an element of $\mathcal{O}[[T]]$, given by $S_x[T] = \Sigma D_x^{[i]} T^i$, i.e., if $p \in P(M,N)$, $S_x[T](p) = S_x(p) \in P(M, N\otimes A[T])$. We have already seen that $S_{x_1+x_2} = S_{x_1} S_{x_2}$, so A-linearity will follow if $S_{\lambda x} = \lambda S_x$; i.e. if $S_{\lambda x}[T] = S_x[\lambda T]$. But this follows immediately from the definition of S_x and the fact that $\lambda x \otimes T = x \otimes \lambda T$.

It remains to show that S_x is of exponential type. Since $S_x[0]$ is clearly the identity, we have only to verify that $S_x[T_1+T_2] = S_x[T_1]S_x[T_2]$. This is also easy:

$$\begin{aligned} S_x[T_1+T_2](p)[z] &= p[x\otimes T_1 + x\otimes T_2 + z] \\ &= S_x[T_1](p)(x\otimes T_2 + z) \\ &= S_x[T_2]S_x[T_1](p)(z). \quad \square \end{aligned}$$

Proof of Theorem (A5).

We have already observed that composition with ℓ_n induces a map $\mathrm{Hom}_A[\Gamma_n(M), N] \to P_n(M,N)$. Let us first prove that it is injective. Let $A[r]$ denote the polynomial algebra $A[T_1 \ldots T_r]$. If $x_1 \ldots x_r$ are elements of M, write

$$p_{A[r]}(x_1\otimes T_1 + x_2\otimes T_2 + \ldots x_r\otimes T_r) = \sum_{\underline{q}} y_{\underline{q}} T^{\underline{q}} \in N\otimes_A A[r] \quad .$$

Since p has weight n, $y_{\underline{q}} = 0$ if $|\underline{q}| \neq n$. Moreover, in $\Gamma_n(M\otimes A[r]) \cong \Gamma_n(M)\otimes A[r]$, we have $\ell_n(x_1\otimes T_1 + \ldots x_r\otimes T_r) = (x_1\otimes T_1 + \ldots x_r\otimes T_r)^{[n]} = \sum_{|\underline{q}|=n} \underline{x}^{[\underline{q}]} \underline{T}^{\underline{q}}$. Since $p = f\circ\ell_n$, we see that $f(\underline{x}^{[\underline{q}]}) = y_{\underline{q}}$ for all q, and since $\{\underline{x}^{[\underline{q}]}\}$ generated $\Gamma_n(M)$ this proves that p determines f.

For the converse, let p be an element of $P_n(M,N)$. It is apparent from the preceeding paragraph that it suffices to construct a linear map $f:\Gamma_n(M) \to N$ such that $f(\underline{x}^{[\underline{q}]}) = y_{\underline{q}}$ whenever $|\underline{q}| = n$, (using the same notation).

In essence, the previous lemma has done this for us: We have an A-linear map $S:M \to \exp(\mathcal{D})$, and hence by (A1), an A-algebra homomorphism $\tilde{S}:\Gamma_A(M) \to \mathcal{D}$, sending $x^{[q]}$ to $D_x^{[q]}$ for any $x \in M$ and any $q \geq 0$. Since S is an algebra homomorphism, it follows that if $\underline{q} = (q_1,\ldots q_r)$, and $x = (x_1\ldots x_r)$, $\tilde{S}(\underline{x}^{[\underline{q}]}) = \underline{D}_{\underline{x}}^{[\underline{q}]}$. Now any $p \in P(M,N)$ determines an A-linear evaluation map $e_p:\mathcal{D} \to N$, sending any D to $D(p)_A(0)$. Composing with $\tilde{S}:\Gamma_A(M) \to \mathcal{D}$, we obtain an A-linear map $\Gamma_A(M) \to N$ sending $\underline{x}^{[\underline{q}]}$ to $\underline{D}_{\underline{x}}^{[\underline{q}]}(p)_A(0)$. The following calculation, which shows that $\underline{D}_{\underline{x}}^{[\underline{q}]}(p)_A(0)$ is none other than $y^{[\underline{q}]}$, completes the proof of the theorem:

$$p_{A[r]}(x_1 \otimes T_1 + \ldots x_r \otimes T_r + z) = \sum_{\underline{q}} (D_{\underline{x}}^{[\underline{q}]}p)_R(z)\underline{T}^{\underline{q}}$$

Evaluating at $z = 0$, we get:

$$p_{A[r]}(x_1 \otimes T_1 + \ldots x_r \otimes T_r) = \sum_{\underline{q}} (D_{\underline{x}}{}^{[\underline{q}]}p)_R(0)\underline{T}^{q} \text{ . } \square$$

Theorem (A5) has several useful computational corollaries:

(A6) <u>Corollary.</u> Let $M' \overset{f}{\underset{g}{\rightrightarrows}} M \xrightarrow{h} M'' \longrightarrow 0$ be an exact sequence of A-linear maps (i.e. h is the cokernel of f and g). Then for each n and N, we have exact sequences:

$$0 \to P_n(M'',N) \to P_n(M,N) \rightrightarrows P_n(M',N)$$

and

$$\Gamma_n(M') \rightrightarrows \Gamma_n(M) \to \Gamma_n(M'') \to 0 \quad .$$

<u>Proof.</u> Since $\otimes R$ is right exact, each $M'\otimes R \rightrightarrows M\otimes R \to M''\otimes R \to 0$ is exact, and the statement about the P_n's follows from the definition. Theorem (A5) then implies the statement about the Γ_n's .

(A7) <u>Corollary.</u> *Suppose* $0 \to K \to M \xrightarrow{g} N \to 0$ *is an exact* sequence of A-modules. Then $\ker \Gamma(g):\Gamma(M) \to \Gamma(N)$ is the ideal generated by $\{x^{[n]} : x \in K , n > 0\}$. □

<u>Proof.</u> From the previous result it is clear that $\operatorname{Ker}\Gamma_n(g)$ is the submodule generated by $\{\underline{x}^{[\underline{q}]}:|\underline{q}| = n, \underline{x} = (x_1 \ldots x_r) \in K^r, r > 0\}$. The corollary follows. □

(A8) <u>Warning:</u> If M is an A/I-module, it is not true that $\Gamma_A(M) \cong \Gamma_{A/I}(M)$, in general. For example, let k be a ring of characteristic $p > 0$, let $A = k[T]$, and let $I = (T)$, $M \cong A/I$, and $M' \cong A/T^pA$. For any A-algebra R, the Frobenius $F_R:R \to R$ induces a map $R/TR \to R/T^pR$, which we can regard as a map $\varphi_R:M\otimes_A R \to M'\otimes_A R$. It is clear that this defines a polynomial function φ: $M \to M'$ of weight p. Therefore there is a linear f: $\Gamma_{A,p}(M) \to M'$ sending $x^{[p]}$ to x' where x and x' are bases of M and M', respectively. Since $Tx' \neq 0$, $Tx^{[p]} \neq 0$, although of course $T\Gamma_{A/I}(M) = 0$. In fact, the reader can see that $\Gamma_{A,p}(M) \cong A/T^pA$, while $\Gamma_{A/I,p}(M) \cong A/TA$. □

We are at last within reach of our goal.

(A9) <u>Theorem.</u> The ideal $\Gamma_A^+(M) \subseteq \Gamma_A(M)$ has a unique PD structure γ such that $\gamma_i(x^{[1]}) = x^{[i]}$ for all i and all $x \in M$.

<u>Proof.</u> Uniqueness is easy. For existence, we consider several cases:

<u>Case 0.</u> A is a $\mathbb{Q}$-algebra.

In this case, $\Gamma_A^+(M)$ has a unique PD structure, given by $\gamma_i(y) = y^i/i!$ *for every* $y \in \Gamma_A^+(M)$. *We need only verify that* $x^{[i]} = (x^{[1]})^i/i!$ if $x \in M$. This is a consequence of the definition, because in $\Gamma_A(M)$ we have, from the defining relations of type III, $xx^{[n]} = (n+1)x^{[n+1]}$, *and hence by induction* $x^n = n!x^{[n]}$.

<u>Case 1.</u> The map $A \to A \otimes_{\mathbb{Z}} \mathbb{Q}$ is injective and M is free.

Let $A' = A\otimes_{\mathbb{Z}}\mathbb{Q}$ and $M' = M\otimes_A A'$. We see from (A3) that $\Gamma_A(M)$ is a free A-module, and that the map $\Gamma_A(M) \to \Gamma_{A'}(M')$ is injective. Now $\Gamma_A(M)$ is the A-subalgebra of $\Gamma_{A'}(M')$ generated by $\{x^{[n]} : x \in M\}$, so that Lemma (3.8) tells us that it is enough to check that $\gamma_i(x^{[n]}) \in \Gamma_A(M)$ for all $i \geq 0$. But in $\Gamma_{A'}(M')$, $\gamma_i(x^{[n]}) = (1/i!)(x^{[n]})^i = (1/i!)[(x^{[1]})^n/n!]^i =$

$$= \frac{(ni)!}{i!(n!)^i} x^{[ni]} = C_{i,n}\, x^{[ni]} .$$

This is an element of $\Gamma_A(M)$ because $C_{i,n}$ is an <u>integer.</u>

<u>Case 2.</u> The map $A \to A\otimes Q$ is injective and M is arbitrary. Find an exact sequence $0 \to K \to F \xrightarrow{f} M \to 0$, with F free. It suffices to show that $\operatorname{Ker} \Gamma_A(f)$ is a sub PD-ideal of $\Gamma_A^+(F)$. By Lemma (3.7) and (A7), we need only check that $\gamma_i(x^{[n]}) \in \operatorname{Ker} \Gamma_A(f)$ if $n \geq 1$, $i \geq 1$, and $x \in K$. But just as before, $\gamma_i(x^{[n]}) = C_{i,n}x^{[ni]} \in \operatorname{Ker} \Gamma_A(f)$. □

<u>Case 3.</u> A and M are arbitrary.

Find a surjective map $P \to A$ such that $P \to P\otimes\mathbb{Q}$ is injective, and let I be the kernel of this map. Then if M is an A-module, we have by (A3) that $\Gamma_A(M) \cong \Gamma_P(M)\otimes_P A$, and hence there is an exact sequence: $\Gamma_P(M)\otimes I \to \Gamma_P(M) \to \Gamma_A(M) \to 0$. We must show that $I\Gamma_P(M) \cap \Gamma_P^+(M)$ is a sub PD-ideal of $\Gamma_P^+(M)$. Clearly $I\Gamma_P(M) \cap \Gamma_P^+(M) = I\Gamma_P^+(M)$ is generated by the set of elements of the form αx, where $\alpha \in I$ and $x \in \Gamma_P^+(M)$. Again by Lemma (3.7), we see that it suffices to check the effect of γ_i on these elements. But $\gamma_i(\alpha x) = \alpha^i\gamma_i(x)$ is still an element of $I\Gamma_P^+(M)$, and so the theorem is proved. □

We can now easily establish the last universal mapping property of $\Gamma_A(M)$, namely, Theorem (3.9) of the text. For the reader's convenience, we repeat its statement:

(3.9) <u>Theorem.</u> Let $\varphi: M \to \Gamma_A^+(M)$ by $x \longmapsto x^{[1]}$. Then if (B,J,δ) is any A-PD-algebra and $\psi: M \to J$ an A-linear map, there is a unique PD morphism $\bar{\psi}: (\Gamma_A(M), \Gamma_A^+(M), \gamma) \longrightarrow (B,J,\delta)$ such that $\bar{\psi}\circ\varphi = \psi$.

Proof. Actually the hard part of this theorem is the establishment of the existence of γ, which we have already done. It is easy to see that if $x \in M$, $\tilde{\psi}(x) = \sum_0^\infty \delta_i(\psi(x))T^i$ belongs to $\exp(B)$, and that $\tilde{\psi}: M \to \exp(B)$ is A-linear. Then by (A1), there is a (unique) A-algebra map $\bar{\psi}: \Gamma_A(M) \to B$ sending $x^{[n]}$ to $\delta_n(\psi(x))$. To see that $\bar{\psi}$ is a PD morphism, it suffices to check elements of the form $x^{[n]}$, since these generated $\Gamma_A^+(M)$ as an ideal. But $\bar{\psi}(\delta_i(x^{[n]})) = \bar{\psi}(C_{i,n}x^{[ni]}) = C_{i,n}\bar{\psi}(x^{[ni]}) = C_{i,n}\delta_{ni}\psi(x) = \delta_i(\delta_n\psi(x)) = \delta_i(\bar{\psi}(x^{[n]}))$. Finally, it is clear that $\bar{\psi}$ is unique, because $\varphi(M) = \Gamma_1(M)$ generates $\Gamma_A(M)$ as an A-PD-algebra. □

There is one more fact that I would like to explain, namely, the relationship between Γ_n and S^n:

(A10) Proposition. *Let M be an A-module, and let $\check{}$ denote $\mathrm{Hom}_A(\ ,A)$. There is a natural A-linear map $S^n(M^\vee) \to \Gamma_n(M)^\vee$ for every n. This map is an isomorphism if M is projective and of finite rank.*

Proof. If $\varphi \in M^\vee$, φ defines a polynomial function weight 1 : $M \to A$ (by the rule $\varphi_R: M\otimes R \to A\otimes R = \varphi\otimes \mathrm{id}_R$). If $\varphi_1 \ldots \varphi_n$ are elements of $M^\vee$, we can multiply these polynomial functions (using multiplication in A) to obtain a polynomial function $\tilde{\varphi}: M \to A$ of weight n. Explicitly, $\tilde{\varphi}_R$ is the composite:

$$\tilde{\varphi}_R:\quad M\otimes R \xrightarrow{(\varphi_{1R},\ldots,\varphi_{nR})} R \times R \ldots \times R \xrightarrow{\text{mult.}} R\ .$$

It is clear that this is compatible with change of R, and has weight n. It is also clear that $\tilde{\varphi}$ depends only on the image φ of $(\varphi_1 \ldots \varphi_n)$ in $S^n(M)$. By Theorem (A5), we see that there is a unique A-linear map $\bar{\varphi}: \Gamma_n(M) \to A$ such that $\bar{\varphi} \circ \ell_n = \tilde{\varphi}$, i.e., such that $\bar{\varphi}(x^{[n]}) = \varphi_1(x) \ldots \varphi_n(x)$ for all $x \in M \otimes R$ and all R. One sees immediately (using the uniqueness) that the map: $S^n(M^\vee) \to \Gamma_n(M)^\vee$ which we have just defined is A-linear.

To prove that we get an isomorphism if M is projective and of finite rank, we may assume that M is free, because both sides are compatible with localization. Suppose $\{x_i : i = 1 \ldots d\}$ is a basis for M and $\{\varphi_i : i = 1 \ldots d\}$ is the dual basis for $M^\vee$. Then $\{\underline{x}^{[\underline{p}]} = x_1^{[p_1]} \ldots x_d^{[p_d]} : |\underline{p}| = n\}$ is a basis for $\Gamma_n(M)$ and $\{\underline{\varphi}^{\underline{q}} = \varphi_1^{q_1} \ldots \varphi_d^{q_d} : |\underline{q}| = n\}$ is a basis for $S^n(M^\vee)$. Recall from the proof of (A5) that $\bar{\varphi}(\underline{x}^{[\underline{p}]})$ is the coefficient of $\underline{T}^{\underline{p}}$ in $\tilde{\varphi}(x_1 \otimes T_1 + \ldots x_d \otimes T_d) \in A[T_1 \ldots T_d]$. This is easy to compute:

$$\tilde{\underline{\varphi}}^{\underline{q}}(x_1 \otimes T_1 + \ldots x_d \otimes T_d) = \prod_{i=1}^{d} \varphi_i^{q_i}(x_1 \otimes T_1 + \ldots x_d \otimes T_d)$$

$$= \prod_{i=1}^{d} T_i^{q_i} = \underline{T}^{\underline{q}} .$$

In other words $\bar{\underline{\varphi}}^{\underline{q}}(\underline{x}^{[\underline{p}]}) = \delta_{p,q}$, so that $\{\underline{\varphi}^{\underline{q}} : |\underline{q}| = n\}$ maps into the dual basis for $\Gamma_n(M)$, as desired. □

This completes the proof of the theorem, but it is of some interest to write down the explicit formula for the pairing between $S^n(M)$ and $\Gamma_n(M)$ we have constructed, for an arbitrary A-module M. It suffices to specify $(\varphi_1 \ldots \varphi_n, x_1^{[q_1]} \ldots x_r^{[q_r]})$, where $\Sigma q_i = n$, where $\varphi_i \in M^\vee$, and where $x_i \in M$. According to the description above, this is just the coefficient of $\underline{T}^{\underline{q}}$ in $\prod_{i=1}^{n} \varphi_i(x_1 \otimes T_1 + \ldots x_r \otimes T_r) \in A[T_1 \ldots T_r]$. This is easily seen to yield the following:

(A11) <u>Remark.</u> $(\varphi_1 \ldots \varphi_n, \underline{x}^{[\underline{q}]}) = \sum_{\alpha} \varphi_{\alpha_1}(x_1) \ldots \varphi_{\alpha_r}(x_r)$, where α ranges over the set of all partitions of $\{1 \ldots n\}$ into r subsets $\alpha_1 \ldots \alpha_r$, where $\mathrm{card}(\alpha_i) = q_i$, and where $\varphi_{\alpha_i}(x_i)$ means $\prod_{j \in \alpha_i} \varphi_j(x_i)$.

REFERENCES

Roby, N. "Lois polynômes et lois formelles en théorie des modules" Ann. de l'E.N.S. 3e sér. 80 (1963) 213-348.

Roby, N. "Les algèbres à puissances divisées" Bull. Sci. Math. Fr. 2e sér 89 (1965) 75-91.

Appendix B

Finiteness of $\mathbb{R}\varprojlim$

In this appendix we will prove the finiteness theorem for $\mathbb{R}\varprojlim$ which we need in §7 in the proof of finiteness of P-adic crystalline cohomology. It seems to be quite essential to work with derived categories. If A is a ring, let $K(A)$ denote the category of complexes of A-modules, and let $K^+(A)$ (resp. $K^-(A)$, resp. $K^b(A)$) the full subcategory consisting of complexes which are bounded below (resp. above, resp. in both directions.) Let $D(A)$ denote the derived category of A-modules (formed by inverting quasi-isomorphisms in the homotopy category of $K(A)$), and $D^+(A)$... the full subcategory consisting of objects which can be represented by objects of $K^+(A)$... . More generally, if X is a site and A is a sheaf of rings on X, $D(X,A)$ will be the derived category of sheaves of A-modules...

The site $\mathbb{N}$ is especially simple, and we can make the construction $\mathbb{R}\Gamma(\mathbb{N},\ )$ rather explicit.

(B1) Remarks. *Let $A.$ be a sheaf of rings on $\mathbb{N}$.*

(B1.1) *A morphism of complexes $A^\cdot \to B^\cdot$ in $K(\mathbb{N},A.)$ is a quasi-isomorphism iff each morphism $A^\cdot_n \to B^\cdot_n$ is.* It follows from this that $D \mapsto D_n$ gives a functor: $D(\mathbb{N},A.) \to D(A_n)$.

(B.1.2) An object D of $D(\mathbb{N},A.)$ lies in $D^+(\mathbb{N},A.)$ (resp. $D^-(\mathbb{N}, A.)$...) iff there is an i such that $H^j(D_n) = 0$ for all $j \geq i$ and all n. (resp...). An object D of $D^-(\mathbb{N},A.)$ has finite tor-dimension iff there exist a and b such that for all n and all A_n-modules M, $H^i(D_n \overset{\mathbb{L}}{\otimes}_{A_n} M) = 0$ unless $i \in [a,b]$.

(B.1.3) The derived functor of $\varprojlim = \Gamma(\mathbb{N},\ )$ is defined on all of $D(\mathbb{N},A.)$, not just on $D^+(\mathbb{N},A.)$. This is a consequence of [RD I 5.3Y] and the well-known fact that $\mathbb{N}$ has finite cohomological dimension. We shall explain this explicitly in the next few remarks.

(B.1.4) A sheaf F on $\mathbb{N}$ is flasque iff the inverse system F is strict, i.e., iff all the maps $F_n \to F_{n-1}$ are surjective. If $F \to G$ is an epimorphism and F is flasque, so is G.

(B.1.5) If $F^\cdot$ is an exact complex of flasque sheaves (not necessarily bounded), then $\Gamma(\mathbb{N},F^\cdot)$ is also exact. If $F^\cdot \to G^\cdot$ is a quasi-isomorphism of complexes of flasque sheaves, $\Gamma(\mathbb{N},F^\cdot) \to \Gamma(\mathbb{N},G^\cdot)$ is a quasi-isomorphism.

(B.1.6) If F is a sheaf of $A.$-modules and if $[F]$ is the sheaf of discontinuous sections of F (i.e. $[F]_n = \prod_{m \leq n} F_m$), then $[F] \to [F]/F \to 0 \to \ldots$ is a flasque resolution of F. Any complex K° of abelian sheaves is quasi-isomorphic to a flasque complex $F^\cdot$, which may be taken to be bounded above (resp. below, resp. in both directions) if $K^\cdot$ is. The image of $\Gamma(\mathbb{N},F^\cdot)$ in

$D(\Gamma(A.))$ depends only on the image of $F^{\bullet}$ in $D(\mathbb{N},A.)$. If $D \in D(\mathbb{N},A.)$, $\mathbb{R}\varprojlim D = \mathbb{R}\Gamma(D)$ is defined to be this image, for any flasque representative $F^{\bullet}$ of D. □

Suppose now that A is a noetherian ring, J-adically separated and complete. Let $A.$ denote the sheaf of rings on $\mathbb{N}$ given by $A_n = A/J^{n+1}$ for all n. Then $A = \Gamma(\mathbb{N},A.) = \mathbb{R}\Gamma(\mathbb{N},A.)$.

We shall soon need the following technical result concerning the behavior of $\mathbb{R}\varprojlim$ and tensor products.

(B2) <u>Proposition.</u> Suppose $D \in K^{-}(\mathbb{N},A.)$. Then:

(B2.1) There exist an $F \in K^{-}(\mathbb{N},A.)$ and a surjective quasi-isomorphism $F \to D$ such that each F_n^q is a projective A_n^q-module and each map $F_n^q \to F_{n-1}^q$ is surjective.

(B2.2) If $F \in K^{-}(\mathbb{N},A.)$ is flat and flasque, $\varprojlim F.$ is flat, and for any finitely generated A-module M, the natural map: $(\varprojlim F.)\otimes_A M \to \varprojlim (F.\otimes_A M)$ is an isomorphism.

(B2.3) If $N_0 \in D^{-}(A_0)$, there is a natural map:

$$\mathbb{R}\varprojlim(D) \overset{\mathbf{L}}{\underset{A}{\otimes}} N_0 \to \mathbb{R}\varprojlim(D \overset{\mathbf{L}}{\underset{A.}{\otimes}} N_0) ,$$

which is an isomorphism if N_0 has finitely generated cohomology.

<u>Proof.</u> To prove the first statement, we first remark that if $K_{n-1}^{\bullet}$ is an acyclic complex of projective A_{n-1}-modules and

is bounded above, then there exists an acyclic complex $K_n^\bullet$ of projective A_n-modules, still bounded above, and a surjective map $K_n^\bullet \to K_{n-1}^\bullet$. In fact, one sees easily by descending induction that the complex $K_{n-1}^\bullet$ splits: $K_{n-1}^q = P_{n-1}^{q-1} \oplus P_{n-1}^q$, with the obvious boundary maps. Now write P_{n-1}^q as a quotient of a free A_n-module P_n^q and take $K_n^q = P_n^{q-1} \oplus P_n^q$ with the obvious boundaries. (As a matter of fact, it is even true that $K_{n-1}^\bullet$ _lifts_ to A_n, as Houzel shows [SGA 5 Exp. XV], but we shall not need this result.)

We prove (B2.1) by constructing the complex $F_n^\bullet$ inductively. Given $F_{n-1}^\bullet \to D_{n-1}^\bullet$, find a $P_n^\bullet \in K^-(A_n)$ consisting of projective A_n-modules and a surjective quasi-isomorphism $P_n^\bullet \to D_n^\bullet$, and then a morphism $P_n^\bullet \to F_{n-1}^\bullet$ covering the given $D_n^\bullet \to D_{n-1}^\bullet$. By the previous paragraph, we can find a surjective map $K_n^\bullet \to K_{n-1}^\bullet$, where $K_{n-1}^\bullet$ is the mapping cone of the identity endomorphism of $F_{n-1}^\bullet$, with $K_n^\bullet$ projective, acyclic, and bounded above. Then set $F_n^\bullet = P_n^\bullet \oplus K_n^\bullet[-1]$, and take the obvious surjection $F_n^\bullet \to F_{n-1}^\bullet$.

To prove (B2.2), we can work with each term of the given complex F, i.e. we may take F. to be a single A.-module. By a well known argument, it suffices to prove that the functor on finitely generated A-modules:

$$\underline{F}: \ M \mapsto \varprojlim (F.\otimes M)$$

is exact.

First we note that the inverse system $\mathrm{Tor}_1^A(F.,M)$ is essentially zero. Note that $\mathrm{Tor}_1^A(F_n,M) = \mathrm{Tor}_1^A(A_n,M) \otimes_{A_n} F_n$

(since F_n is A_n-flat), so it suffices to prove that the inverse system $T(M) = \mathrm{Tor}_1^A(A.,M)$ is essentially zero. Now we have a natural inclusion: $T_n(M) \subseteq J^{n+1}\otimes M$, and we can apply Artin-Rees to find an integer ν such that $J^{m+\nu}(J^{n+1}\otimes M) \cap T_n(M) \subseteq J^m T_n(M)$ for all $m \geq 0$. Since $J^{n+1}T_n(M) = 0$, this implies that $J^m(J^{n+1}\otimes M) \cap T_n(M) = 0$ for $m > n+\nu$. But the image of $T_{n+m}(M) \to T_n(M)$ is contained in this intersection, and so the arrow vanishes for $m >> 0$.

The exactness of the functor $\underline{F}$ follows easily: If $0 \to M' \to M \to M'' \to 0$ is exact, we obtain for each n an exact sequence:

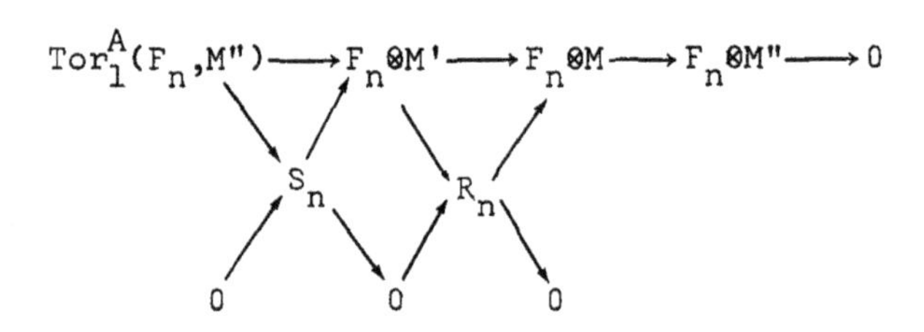

Since $F.$ is flasque, so is $F.\otimes M'$, hence also $R.$, and since $\mathrm{Tor}_1^A(F.,M'')$ is essentially zero, so is $S.$. One deduces immediately that the sequence

$$0 \to \varprojlim(F.\otimes M') \to \varprojlim(F.\otimes M) \to \varprojlim(F.\otimes M'') \to 0$$

is exact.

Before we prove (B2.3), let us mention as a corollary to (B2.2) that if E is a flat A-module, $\hat{E}$ is again flat, and $\hat{E}\otimes A_n \cong E\otimes A_n$.

Suppose $D \in D^-(\mathbb{N},A.)$, and let $F. \simeq D$ be as in (B2.1). Then the complex $F = \varprojlim F.$ consists of flat A-modules, and is isomorphic to $\mathbb{R}\varprojlim D$, so that we have: $F \otimes_A N. \simeq \mathbb{R}\varprojlim D \overset{\mathbb{L}}{\underset{A}{\otimes}} N_0$.

On the other hand, $F.$ is a flat A.-module, so $F.\otimes_{A.} N_0 \simeq D \overset{\mathbb{L}}{\underset{A.}{\otimes}} N_0$, and since $F.\otimes_{A.} N_0$ is still flasque, $\varprojlim(F.\otimes_{A.} N_0) \simeq \mathbb{R}\varprojlim(D \overset{\mathbb{L}}{\underset{A}{\otimes}} N_0)$. Thus, the natural map:

$$(\varprojlim F.)\otimes_A N_0 \to \varprojlim(F.\otimes_A N_0) \xrightarrow{\cong} \varprojlim(F.\otimes_{A.} N_0)$$

provides us with the arrow of (B2.3). If, moreover N_0 has finitely generated cohomology, it is isomorphic to a complex M of finitely generated A_0-modules, and since each term of the complex $F.\otimes M$ will involve a uniformly finite direct sum of terms as in (B2.2), we see that the arrow is an isomorphism. □

Our aim is to compare the categories $D(\mathbb{N},A.)$ and $D(A)$. We have defined a functor $\mathbb{R}\varprojlim$: $D(\mathbb{N},A.) \to D(A)$, and there is an obvious functor back. In fancy terms, there is a morphism f: $(\tilde{\mathbb{N}},A.) \to (pt.,A)$ of ringed topoi given by Γ, and we can form $\mathbb{L}f^*$ as described in §7. Explicitly:

B.3 <u>Definition.</u> If M is an A-module, let $S(M)_n = M\otimes_A A_n$, a sheaf of A.-modules on $\mathbb{N}$. Then let $\mathbb{L}S$: $D^-(A) \to D^-(\mathbb{N},A.)$ denote the left derived functors of S.

It is clear that $\mathbb{L}S$ can be computed using flat resolutions. Moreover, $\mathbb{L}S(M)_n = M \overset{\mathbb{L}}{\underset{A}{\otimes}} A_n$.

B.4 Definition. An object D of $D(\mathbb{N},A.)$ is "quasi-consistent" iff the natural maps $D_n \overset{\mathbf{L}}{\otimes}_{A_n} A_{n-1} \to D_{n-1}$ are all isomorphisms. (For this to make sense, we must have $D_n \in D^-(A_n)$ for all n.) We say D is "consistent" iff it is quasi-consistent and each D_n has finitely generated cohomology.

B.5 Proposition. If $D \in D^-(\mathbb{N},A)$, the following are equivalent:

1) D is quasi-consistent.
2) D is isomorphic to $\mathbf{L}S(M)$ for some $M \in D^-(A)$.
3) The natural map $\mathbf{L}S\,\mathbb{R}\varprojlim D \to D$ is an isomorphism.

Proof. Let us begin by defining the arrow in (3). Choose a flasque representative $D^{\bullet}$ of D in $K^-(\mathbb{N},A.)$, so that $\varprojlim D_n^{\bullet}$ represents $\mathbb{R}\varprojlim D$. Choose a projective resolution $L^{\bullet} \to \varprojlim D_n^{\bullet}$. Then for each n, we have a commutative diagram:

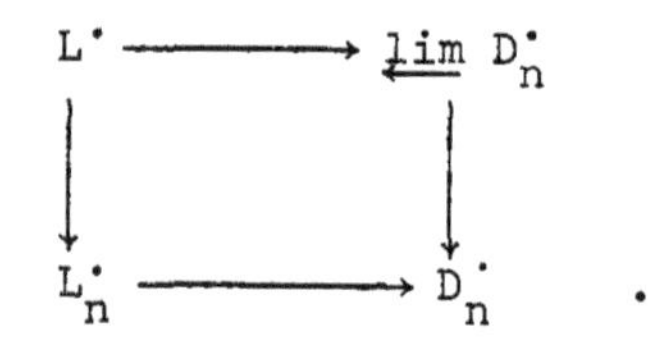

This defines a morphism of complexes of sheaves $S(L^{\bullet}) \to D^{\bullet}$, hence a morphism of functors $\mathbf{L}S \circ \mathbb{R}\varprojlim \to \mathrm{id}$ on $D^-(\mathbb{N},A.)$.

The implication (3) ⇒ (2) is obvious. To prove that (2) ⇒ (1), we are reduced to checking that the map $[M \overset{\mathbf{L}}{\otimes}_A A_n] \overset{\mathbf{L}}{\otimes}_{A_n} A_{n-1} \longrightarrow M \overset{\mathbf{L}}{\otimes}_A A_{n-1}$ is an isomorphism. Taking a flat complex representing M, this is obvious. To prove that (1) ⇒ (3), we are reduced to:

B.5.1 Claim. If $D \in D^{-}(\mathbb{N},A.)$ is quasi-consistent, the natural map $(\mathbb{R}\varprojlim D) \overset{\mathbb{L}}{\underset{A}{\otimes}} A_n \to D_n$ is an isomorphism, for every n.

Proof. A straightforward extension of (B2.3) implies that there is a natural isomorphism:

$$(\mathbb{R} \varprojlim D) \overset{\mathbf{L}}{\underset{A}{\otimes}} A_n \longrightarrow \mathbb{R}\varprojlim_{m \geq n} (D_m \overset{\mathbb{L}}{\underset{A_m}{\otimes}} A_n) \quad ,$$

and so it suffices to find a natural isomorphism:

$$\mathbb{R}\varprojlim_{m \geq n} (D_m \overset{\mathbf{L}}{\underset{A_m}{\otimes}} A_n) \longrightarrow D_n \quad .$$

Choose $F. \simeq D$ as in (B2.1); then $D_m \overset{\mathbb{L}}{\underset{A_m}{\otimes}} A_n \simeq F_m \otimes_{A_m} A_n$, and since the system is flasque, $\mathbb{R}\varprojlim D_m \overset{\mathbf{L}}{\underset{A_m}{\otimes}} A_n \simeq \varprojlim (F_m \otimes_{A_m} A_n)$. But each map of complexes $F_m \otimes_{A_m} A_n \to F_n$ is a quasi-isomorphism, and it follows that the same is true in the limit.

(B.6) Lemma. Suppose that $D \in D(\mathbb{N},A.)$ is quasi-consistent. Then:

(B.6.1) In fact $D \in D^{-}(\mathbb{N},A.)$.

(B.6.2) If $D_0 \in D^{-}_{ft}(A_0)$, then each $D_n \in D^{-}_{ft}(A_n)$.

(B.6.3) If D_0 has bounded tor-dimension, so do D and $\mathbb{R}\varprojlim D$.

Proof. Since $D_0 \in D^{-}(A_0)$, it is (quasi)-isomorphic to a $K_0^{\bullet}$ of flat A_0-modules with $K_0^i = 0$ if $i > n_0$, and we see that $H^i(D_0 \overset{\mathbb{L}}{\otimes}_{A_0} M_0) \cong H^i(K_0^{\cdot} \otimes_{A_0} M_0) = 0$ if $i > n_0$, for every A_0-module M_0. In particular, $0 = H^i(D_0 \overset{\mathbb{L}}{\otimes}_{A_0} J^n/J^{n+1}) \cong$ $H^i(D_n \overset{\mathbb{L}}{\otimes}_{A_n} A_0 \overset{\mathbb{L}}{\otimes}_{A_0} J^n/J^{n+1}) = H^i(D_n \overset{\mathbb{L}}{\otimes}_{A_n} J^n/J^{n+1})$ for $i > n_0$. From the long exact sequence:

$$\to H^i(D_n \overset{\mathbb{L}}{\otimes}_{A_n} J^n/J^{n+1}) \to H^i(D_n) \to H^i(D_{n-1})$$

and induction, we see that $H^i(D_n) = 0$ for $i > n_0$, so that $D \in D^-(\mathbb{N}, A.)$. This proves (B6.1); the proofs of (B6.2) and the first part of (B6.3) are similar.

Let us verify that $\mathbb{R}\varprojlim D$ has finite tor-dimension if D_0 does. Choose a flat and flasque $F. \simeq D$ as in (B2.1), so that $F = \varprojlim F.$ is $\mathbb{R}\varprojlim D$; since it is still flat, $F \otimes M \simeq \mathbb{R}\varprojlim D \overset{\mathbb{L}}{\otimes}_A M$ for an arbitrary A-module M. We must show that there exists an m such that $H^i(F \otimes M) = 0$ for all $i < m$ and all M, and we can assume that M is finitely generated. Then (B2.2) tells us that $F \otimes_A M \cong \varprojlim (F. \otimes_A M) \cong \varprojlim (F. \otimes_{A.} (M \otimes_A A.))$. But (B6.3) says that D. has finite tor-dimension, so there exists an m such that $H^i(D_n \overset{\mathbb{L}}{\otimes}_{A_n} (M \otimes_A A_n)) \cong H^i(F_n \otimes_A M) = 0$ for all $i < m$ and all n. Since the complex $F. \otimes M$ is flasque, there is an exact sequence:

$$0 \to R^1\varprojlim H^{i-1}(F. \otimes M) \to H^i(\varprojlim F. \otimes M) \to \varprojlim H^i(F. \otimes M) \to 0 ,$$

and hence $H^i(F \otimes M) = 0$ for $i < m$. □

We now come to the main finiteness result:

B.7 <u>Proposition.</u> Suppose $D \in D(\mathbb{N},A.)$ is quasi-consistent and $D_0 \in D^-_{ft}(\mathbb{N},A)$. Then:

B.7.1 The inverse systems $H^i(D.)$ satisfy ML.

B.7.2 The natural maps $H^i(\mathbb{R}\varprojlim D) \longrightarrow \varprojlim H^i(D.)$ are isomorphisms.

B.7.3 $\mathbb{R}\varprojlim D \in D^-_{ft}(A)$.

<u>Proof.</u> By the previous lemma $D \in D^-(\mathbb{N},A)$. It is quite easy to prove the proposition if A is a DVR with uniformizing parameter π. Then (1) is automatic, since the cohomology modules $H^i(D_n)$ are Artinian. (2) follows, and then one concludes easily that $H^i(\mathbb{R}\varprojlim D)$ is π-adically separated and complete. Since we have $\mathbb{R}\varprojlim D \overset{\mathbb{L}}{\otimes} A_0 \cong D_0$, $H^i(\mathbb{R}\varprojlim D)\otimes_A A_0 \xrightarrow{\cong} H^i(D_0)$, hence is finitely generated, hence so is $H^i(\mathbb{R}\varprojlim D)$.

The general proof is a bit more complicated. Begin by applying (B.5), which allows us to replace D by $\mathbb{L}S(M)$, where $M = \mathbb{R}\varprojlim D$. Since M lies in $D^-(A)$ by (B.6.1) and (B.1.6), we can choose a flat complex $L^\cdot$ representing M. Then the flasque complex $S(L^\cdot)$ represents $\mathbb{L}S(M)$, hence $\varprojlim S(L^\cdot)$ represents $\mathbb{R}\varprojlim \mathbb{L}\, S(L^\cdot) \cong \mathbb{R}\varprojlim D \cong M$. In particular, the natural map $L^\cdot \to \varprojlim L^\cdot \otimes A_n$ is a quasi-isomorphism.

Moreover, since $D_0 \in D^-_{ft}(A_0)$, we can choose a complex $D^\cdot_0$ of finitely generated projective A_0-modules representing D_0, and if B_0 is a noetherian A_0-algebra, $D^\cdot_0 \otimes_{A_0} B_0$ represents $D_0 \overset{\mathbb{L}}{\otimes}_{A_0} B_0$, so that $H^i(D^\cdot_0 \otimes_{A_0} B_0)$ is finitely generated, for all i. Since $L^\cdot$

is flat, $gr_J L^\bullet \cong gr_J A \otimes_{A_0} L_0^\bullet$, so that, in particular, $H^i(gr_J L^\bullet)$ is a finitely generated $gr_J(A)$-module, for all i. This proves that $L^\bullet$ satisfies the hypotheses of the following Lemma, which therefore will prove the Proposition.

B.8 Lemma. Suppose $L^\bullet$ is a complex of flat A-modules such that

a) The natural map $L^\bullet \to \varprojlim L^\bullet \otimes A_n$ is a quasi-isomorphism.

b) Each $H^i(gr_J L^\bullet)$ is a finitely generated $gr_J A$-module.

Then:

B.8.1 The inverse systems $H^i(L^\bullet \otimes A_n)$ satisfy M.L.

B.8.2 The natural maps $H^i(L^\bullet) \to \varprojlim H^i(L^\bullet \otimes A_n)$ are isomorphisms.

B.8.3 Each $H^i(L^\bullet)$ is a finitely generated A-module.

Proof. This follows, I believe, from [EGA 0_{III} 13.7.7]. However, I prefer to give a slightly different proof. Let us begin by recalling the following easy result, of which the above is "the derived category version".

Step 0: Suppose $F^\bullet$ is a filtration on the A-module M which is compatible with the J-adic filtration, i.e. $J^k F^\nu M \subseteq F^{\nu+k} M$ for all k and ν. Then:

(a) $gr_F M$ has a natural $gr_J A$-module structure.

(b) If $gr_F M$ is generated as a $gr_J A$-module in degrees $\le \nu$, then $F^{\nu+k} M \subseteq J^k F^\nu M + F^m M$ for all k and $m \ge 0$.

(c) If $gr_F M$ is finitely generated as a $gr_J A$-module and if M is F-adically separated, then M is finitely generated as an A-module.

Fix $n \geq 0$, and consider the filtration $F^{\cdot}$ on the complex $J^nL^{\cdot}$ given by $F^m = J^nL^{\cdot}$ if $m \leq n$, $F^m = J^mL^{\cdot}$ if $m \geq n$. We have a spectral sequence $E_1^i = H^i(gr_F J^nL^{\cdot})$ with abuttment $(H^i(J^nL^{\cdot}), F^{\cdot})$, where $F^mH^i(J^nL^{\cdot}) = \mathrm{Im}[H^i(F^mJ^nL^{\cdot}) \to H^i(J^nL^{\cdot})]$. The terms E_r^i for $r = 1 \ldots \infty$ all are $gr_J A$-modules, and the differentials are $gr_J A$-linear.

<u>Step 1:</u> E_r^i is finitely generated as a $gr_J A$-module if $1 \leq r < \infty$.

<u>Proof.</u> $gr_F J^nL^{\cdot} \longrightarrow gr_F L^{\cdot}$ is (as a group) a direct summand, consisting of those terms of degree $\geq n$. Hence $H^i(gr_F J^nL^{\cdot})$ is a submodule of the finitely generated $gr_J A$-module $H^i(gr_F L^{\cdot})$. This proves that E_1^i is finitely generated, hence so is E_r^i if $1 \leq r < \infty$.

<u>Step 2:</u> $E_r^i = E_\infty^i$ for $r >> 0$, hence E_∞^i is finitely generated.

<u>Proof.</u> For each i, let B_{r+1}^i be the image of d_r^{i-r}: $E_r^{i-r} \to E_r^i$, and let $\bar{B}_{r+1}^i$ be the preimage of B_{r+1}^i in E_2^i Then $\bar{B}_r^i \subseteq \bar{B}_{r+1}^i \subseteq \ldots$ forms an ascending chain of $gr_J A$-submodules, hence is eventually stationary. This implies that $d_r^{i-r} = 0$ for $r >> 0$. Since the same is true for d_r^i, we have Step 2.

<u>Step 3:</u> Let $F^mH^i(L_n^{\cdot}) = \mathrm{Im}[H^i(L_m^{\cdot}) \to H^i(L_n^{\cdot})]$. This filtration is compatible with the J-adic filtration, and $gr_F H^i(L_n^{\cdot})$ is a finitely generated $gr_J A$-module.

<u>Proof.</u> For $m \geq n$, we have an exact sequence:

$0 \to F^mL^{\cdot} \to L^{\cdot} \to L_m^{\cdot} \to 0$, hence a diagram:

$$\begin{array}{ccccccc} H^i(L^\cdot) & \longrightarrow & H^i(L^\cdot_m) & \longrightarrow & H^{i+1}(F^mL^\cdot) & \longrightarrow & H^{i+1}(L^\cdot) \\ \downarrow\cong & & \downarrow\alpha & & \downarrow\beta & & \downarrow\cong \\ H^i(L^\cdot) & \longrightarrow & H^i(L^\cdot_n) & \xrightarrow{\delta} & H^{i+1}(J^nL^\cdot) & \longrightarrow & H^{i+1}(L^\cdot) \end{array}$$

It follows from this that $\mathrm{Im}(\alpha) = \delta^{-1}(\mathrm{Im}(\beta))$; i.e. $\delta^{-1}(F^mH^{i+1}(J^nL^\cdot)) = F^mH^i(L^\cdot_n)$. Hence $gr_FH^i(L^\cdot_n) \hookrightarrow gr_FH^{i+1}(J^nL^\cdot)$, a finitely generated gr_JA-module. The statement follows.

Step 4: The inverse systems $H^i(L^\cdot_n)$ satisfy ML. (This proves (B8.1).)

Proof. Suppose $gr_FH^i(L^\cdot_n)$ is generated in degrees $\leq \nu$. Then by Step 0, $F^{\nu+k}H^i \subseteq J^kF^\nu H^i + F^mH^i$ for all m and $k \geq 0$. Take $k = n+1$; then since $J^{n+1}H^i(L^\cdot_n) = 0$, $F^{\nu+n+1}H^i \subseteq F^mH^i$ for all m — and this is the condition ML.

Step 5: The maps $H^i(L^\cdot) \to \varprojlim H^i(L^\cdot_n)$ are isomorphisms. (This proves (B8.2).)

Proof. Since the maps $L^\cdot_n \to L^\cdot_{n-1}$ are all surjective and since $H^{i-1}(L^\cdot_n)$ satisfies ML, the map $H^i(\varprojlim L^\cdot) \to \varprojlim H^i(L^\cdot)$ is an isomorphism. Since the map $H^i(L^\cdot) \to H^i(\varprojlim L^\cdot)$ is also an isomorphism, we are done.

Step 6: The A-modules $H^i(L^\cdot)$ are finitely generated. (This proves (B8.3).).

Proof. Since $F^nH^i(L^\cdot) = \mathrm{Ker}\,[H^i(L^\cdot) \to H^i(L^\cdot_n)]$, step 5 shows that $H^i(L^\cdot)$ is F-adically separated. Since $gr_FH^i(L^\cdot)$

is finitely generated, we may apply Step 0. This completes the proof of (B.8). □

B.9 Corollary. $\mathbb{R}\varprojlim$ and $\mathbb{L}S$ define an equivalence of categories: $D^-_{con}(\mathbb{N},A.) \rightleftarrows D^-_{ft}(A)$, where $D^-_{con}(\mathbb{N},A.)$ is the full subcategory of $D^-(\mathbb{N},A)$ of consistent complexes (B4).

Proof. If $D \in D^-(\mathbb{N},A.)$ and is only quasi-consistent, the map $\mathbb{L}S\,\mathbb{R}\varprojlim D \to D$ is an isomorphism. If $D \in D^-_{con}(\mathbb{N},A.)$, the previous result implies that $\mathbb{R}\varprojlim D \in D^-_{ft}(A)$. It remains only to prove that $\mathbb{R}\varprojlim \mathbb{L}S(M) \cong M$ if $M \in D^-_{ft}(A)$. Let $M^\bullet$ be a resolution of M by finitely generated free A-modules. Then $S(M^\bullet)$ is flasque, the natural map $M^\bullet \to \varprojlim S(M^\bullet)$ is an isomorphism, and so follows the corollary. □

B.10 Corollary. If $D \in D^-(\mathbb{N},A.)$ is quasi-consistent, $\mathbb{R}\varprojlim D$ is perfect iff D_0 is.

Proof. This follows immediately from the above results and (B6.3).

B.11 Corollary. *There is a natural equivalence of categories between $D_{perf}(A)$ and the category "Pr(A.)" consisting of inverse systems $\{D_n \in D_{perf}(A_n)$ for all $n\}$ such that the transition maps $D_n \to D_{n-1}$ induce isomorphisms $D_n \overset{\mathbb{L}}{\otimes}_{A_n} A_{n-1} \to D_{n-1}$ for all n.*

Proof. If $M \in D_{perf}(A)$, let $P(M)_n = M \overset{\mathbb{L}}{\otimes}_A A_n \in D_{perf}(A_n)$; then P defines a functor: $D_{perf}(A) \to Pr(A.)$.

First let's prove that P is fully faithful. If M and N lie in $D_{perf}(A)$, we can choose bounded complexes $M^\bullet$ and $N^\bullet$ of finitely generated projective A-modules representing M and N . Then $Hom^\bullet_{D(A)}[M,N] = \mathbb{R}\,Hom[M,N] = H^0(H^\bullet)$, where $H^\bullet$ is the complex $Hom^\bullet_A[M^\bullet,N^\bullet]$. This $H^\bullet$ is again a complex of finitely generated flat A-modules, which evidently satisfies the hypotheses of (B.9). Therefore the natural map
$H^0(H^\bullet) \to \varprojlim H^0(H^\bullet \otimes_A A_n)$ is an isomorphism. Moreover,
$H^\bullet \otimes_A A_n \cong Hom_{A_n}[M^\bullet \otimes_A A_n, N^\bullet \otimes_A A_n]$, so that
$H^0(H^\bullet \otimes A_n) \cong Hom_{D(A_n)}[P(M)_n, P(N)_n]$, and
$\varprojlim H^0(H^\bullet \otimes A_n) \cong Hom_{Pr}[P(M), P(N)]$. Thus, P is fully faithful.

To prove that our functor is essentially surjective, suppose $\{D_n : n \in \mathbb{N}\}$ is an object in Pr(A.). For each n, choose a complex of finitely generated projective A_n-modules $D_n^\bullet$ representing D_n . Then the given morphism $D_n \to D_{n-1}$ in $D(A_n)$ comes from a morphism of complexes $D_n^\bullet \to D_{n-1}^\bullet$, unique up to homotopy. Make a choice of such a morphism for each n; then we get an inverse system of complexes $\{D_n^\bullet : n \in \mathbb{N}\}$ – i.e. an element of $K(\mathbb{N},A.)$. Let D be its image in $D(\mathbb{N},A.)$;

By (B6) $D \in D^{-}(\mathbb{N},A.)$ and is consistent, hence by (B.7), $\mathbb{R}\varprojlim D \in D^{-}_{ft}(A)$. It is clear from (B.5.1) that $\mathbb{R}\varprojlim D \overset{\mathbb{L}}{\otimes} A_n \cong D_n$ for all n, and the proof of the corollary is complete. □

BIBLIOGRAPHY

Berthelot, P. - "Cohomologie cristalline des schémas de caractéristique $p > 0$" Lecture Notes in Mathematics N° 407, Springer Verlag (1976).

Bloch, S. - "K-theory and crystalline cohomology" to appear in Publ. Math. I.H.E.S., N° 47 (1977).

Deligne P. - "La conjecture de Weil I" Publ. Math. I.H.E.S. N° 43 (1974).

———— "Théorie de Hodge II" Publ. Math. I.H.E.S. N° 40 (1971) 5-58.

———— "Théorie de Hodge III" Publ. Math. I.H.E.S. N° 44 (1975) 5-77.

Dwork, B. - "On the zeta function of a hypersurface II" Ann. of Math. 80 N° 2 (1964) 227-299.

Grothendieck, A. - "Crystals and the de Rham cohomology of schemes" in Dix exposés sur la Cohomologie des schémas, North Holland (1968).

———— "On the de Rham cohomology of algebraic varieties" Publ. Math. I.H.E.S. N° 29 (1966) 95-103.

Grothendieck, A. and Dieudonné, J. - "Eléments de Géométrie Algébrique" Publ. Math. I.H.E.S. N° 11 (1961).

Grothendieck, A. et al. - "Séminaire de Géométrie Algébrique" Lecture Notes in Mathematics N°s 269, 270, 305, 225, Springer Verlag.

Hartshorne, R. - "On the de Rham Cohomology of Algebraic Varieties" Publ. Math. I.H.E.S. N° 45 (1976) 5-99.

Katz, N. - "Nilpotent Connections and the Monodromy Theorem ..." Publ. Math. I.H.E.S. N° 39 (1970) 175-232.

———— "On the differential equations satisfied by period matrices" Publ. Math. I.H.E.S. N° 35 (1968) 71-106.

Katz, N. and Oda, T. - "On the differentiation of de Rham cohomology classes with respect to parameters" J. Math. Kyoto U. 8 (1968) 199-213.

Kleiman, S. - "Algebraic cycles and the Weil conjectures" in Dix exposés sur la Cohomologie des Schémas, North Holland (1968) 359-386.

Lubkin, S. - "A p-adic proof of the Weil conjectures" Ann. of Math. N° 87 (1968) 105-255.

Manin, Yu. - "The theory of commutative formal groups over fields of finite characteristic" Russian Math. Surv. 18 (1963).

Mazur, B. - "Frobenius and the Hodge Filtration" Bull. A.M.S. 78 (1972) 653-667.

_________ "Frobenius and the Hodge Filtration - estimates" Ann. of Math. 98 (1973) 58-95.

Monsky, P. - "P-adic Analysis and Zeta Functions" Kinokuniya Book Store, Tokyo (1970).

Ogus, A. - "Local Cohomological Dimension of Algebraic Varieties" Ann. of Math. 98 (1973) 327-365.

Roby, N. - "Lois polynômes et lois formelles en théorie des modules" Ann. de l'E.N.S. 3è série 80 (1963) 213-348.

_________ "Les algèbres à puissances divisées" Bull. Sci. Math. Fr. 2è série 89 (1965) 75-91.

Serre, J.P. - "Quelques propriétés des variétés abéliennes en caractéristique p" Am. J. Math. 80 N° 3 (1958) 715-739.

_________ "Sur la topologie des variétés algébriques en caractéristique p" Symp. Int. de Top. Alg., Univ. Nac. Aut. de Mexico (1958) 24-53.

Zariski, O. - "Algebraic Surfaces" (second supplemented edition), Springer Verlag (1971).

Lubkin, S. - "A p-adic proof of the Weil conjectures" Ann. of Math. N° 87 (1968) 105-255.

Manin, Yu. - "The theory of commutative formal groups over fields of finite characteristic" Russian Math. Surv. 18 (1963).

Mazur, B. - "Frobenius and the Hodge Filtration" Bull. A.M.S. 78 (1972) 653-667.

________ "Frobenius and the Hodge Filtration - estimates" Ann. of Math. 98 (1973) 58-95.

Monsky, P. - "P-adic Analysis and Zeta Functions" Kinokuniya Book Store, Tokyo (1970).

Ogus, A. - "Local Cohomological Dimension of Algebraic Varieties" Ann. of Math. 98 (1973) 327-365.

Roby, N. - "Lois polynômes et lois formelles en théorie des modules" Ann. de l'E.N.S. 3è série 80 (1963) 213-348.

________ "Les algèbres à puissances divisées" Bull. Sci. Math. Fr. 2è série 89 (1965) 75-91.

Serre, J.P. - "Quelques propriétés des variétés abéliennes en caractéristique p" Am. J. Math. 80 N° 3 (1958) 715-739.

________ "Sur la topologie des variétés algébriques en caractéristique p" Symp. Int. de Top. Alg., Univ. Nac. Aut. de Mexico (1958) 24-53.

Zariski, O. - "Algebraic Surfaces" (second supplemented edition), Springer Verlag (1971).

Library of Congress Cataloging in Publication Data

Berthelot, Pierre, 1943-
Notes on crystalline cohomology.

(Mathematical notes ; 21)
1. Geometry, Algebraic. 2. Homology theory.
3. Functions, Feta. I. Ogus, Arthur, joint author.
II. Title. III. Title: Crystalline cohomology.
IV. Series: Mathematical notes (Princeton, N. J.)
; 21.
QA564.B46 512'.33 78-57039
ISBN 0-691-08218-9

GPSR Authorized Representative: Easy Access System Europe - Mustamäe tee 50, 10621 Tallinn, Estonia, gpsr.requests@easproject.com

www.ingramcontent.com/pod-product-compliance
Ingram Content Group UK Ltd.
Pitfield, Milton Keynes, MK11 3LW, UK
UKHW021823060425
457147UK00006B/124

* 9 7 8 0 6 9 1 6 2 8 0 8 0 *